Ivan Ricardo Carvalho
Vinícius J. Szareski
Maicon Nardino

Improvement and seed production of annual crops

Ivan Ricardo Carvalho
Vinícius J. Szareski
Maicon Nardino

Improvement and seed production of annual crops

Soya, Maize, Wheat and Beans

ScienciaScripts

Imprint
Any brand names and product names mentioned in this book are subject to trademark, brand or patent protection and are trademarks or registered trademarks of their respective holders. The use of brand names, product names, common names, trade names, product descriptions etc. even without a particular marking in this work is in no way to be construed to mean that such names may be regarded as unrestricted in respect of trademark and brand protection legislation and could thus be used by anyone.

Cover image: www.ingimage.com

This book is a translation from the original published under ISBN 978-613-9-73249-4.

Publisher:
Sciencia Scripts
is a trademark of
Dodo Books Indian Ocean Ltd. and OmniScriptum S.R.L publishing group

120 High Road, East Finchley, London, N2 9ED, United Kingdom
Str. Armeneasca 28/1, office 1, Chisinau MD-2012, Republic of Moldova, Europe
Printed at: see last page
ISBN: 978-620-7-88902-0

Description of the authors

Ivan Ricardo Carvalho: Agricultural Engineer from the Federal University of Santa Maria, Frederico Westphalen campus (2014). Specialist in Seed Science and Technology (2018). Master's degree from the Postgraduate Programme in Agronomy, Agriculture and the Environment at the Federal University of Santa Maria (2015). PhD from the Postgraduate Programme in Agronomy, Plant Genetic Improvement at the Federal University of Pelotas (2018). Post-Doctorate in Seed Science and Technology, he works in the area of plant production and major crops.

Vinícius Jardel Szareski: PhD student in the Postgraduate Programme in Seed Science and Technology at the Federal University of Pelotas (UFPel). Master's in Seed Science and Technology (2017). Agricultural Engineer from the Federal University of Santa Maria UFSM - Frederico Westphalen Campus - RS (2015). Agricultural Technician from Sociedade Educacional Três de Maio (SETREM). He works in the areas of Seed Production and Technology with an emphasis on seed quality control, production and cultivation of forage species, agricultural experimentation, genotype x environment interaction, adaptability and stability, genetic parameters, positioning of genotypes, identification of macro-environments for seed production, abiotic stresses and vigour index in soya and wheat seeds.

Maicon Nardino:Graduated in agronomic engineering from the federal university of santa maria - Frederico Westphalen campus (2011). Master's degree in agronomy, plant breeding and agricultural experimentation from the federal university of santa maria - Frederico Westphalen campus (2013). PhD in Agronomy, Plant Breeding from the Federal University of Pelotas (2015). He is currently a post-doctoral student in experimental statistics and biometrics. He works in the areas of plant breeding, biometrics and plant production.

Francisco Amaral Villela: Degree in Agricultural Engineering from the Federal University of Pelotas/UFPel (1979), specialisation in Physics Teaching from the Catholic University of Pelotas-UCPel (1980), Master's degree in Agronomy from the Federal University of Pelotas-UFPel (1985), PhD in Plant Science from the University of São Paulo-USP (1991) and post-doctorate in the Plant Production Department of the Luiz de Queiroz School of Agriculture-ESALQ/USP (2001). He is currently a Full Professor at the Federal University of Pelotas and a Research Productivity Fellow of the National Council for Scientific and Technological Development (CNPq). FAO consultant in the United Nations Development Programme (UNDP) for Cuba in 2003. Coordinator of the Postgraduate Programme in Seed Science and Technology/UFPel,C and from 08/2012 to 10/2014. Member of the Judging Committee for Productivity Grants in Technological Development and Innovative Extension - Dt (CNPq). Second Vice-President of ABRATES, 2011-2013 term and 2013-2015 term. Head of the CNPq Seed Science and Technology Research Group. Coordinator of the Specialisation Course in Seed Science and Technology/UFPel, from 09/2014 to 09/2016. Has experience in Agronomy, with an emphasis on Seed Science and Technology, working mainly on the following subjects: drying, processing, storage and seed quality control.

Velei Queiroz de Souza:He completed his post-doctorate in agronomy at the Federal University of Pelotas in 2008 - concentrating on plant breeding. He is currently Assistant Dean for Research, Postgraduate Studies and Innovation at the Federal University of Pampa. He works in agronomy, with an emphasis on experimental statistics and genetic improvement. In his professional activities, he has interacted with more than 30 collaborators in co-authoring scientific papers. In his Lattes CV, the most frequent terms used to contextualise his scientific and technological production are: plant breeding, experimental statistics, progeny selection, genotype and environment interaction, and resistance to biotic and abiotic factors.

SUMMARY

CHAPTER 1

Seed production and the implications of Asian rust *(Phakopsora pachyrhizi)* on soya beans

Vinícius Jardel Szareski, Ivan Ricardo Carvalho, Maicon Nardino, Francisco Amaral Villela

Soya is the main oilseed produced and consumed in the world. It is the main source of protein, which is an essential constituent in the manufacture of animal feed and human food, and is also used as a raw material for the extraction of biodiesel in Brazil (BRUMetal., 2005).

Brazil is one of the world's largest soya producers and exporters. Brazilian production reached 86,120.8 million tonnes in the 2013/14 harvest. Rio Grande do Sul is responsible for the 3° largest production of soya beans in the country, behind only Mato Grosso and Paraná (CONAB, 2015).

Despite the growing increase in soya production and productivity, the incidence of diseases is a limiting factor for soya production potential. Asian soya rust is considered the main disease, due to its high damage potential, where the main damage caused by the disease is early defoliation of the plant, preventing the complete formation of grains. The earlier the defoliation occurs, the smaller the grain size and, consequently, the lower the grain yield (MERCHING et al., 1989).

The extent of the damage that the pathogen can cause depends on the moment it enters the crop, the favourable weather conditions and

its multiplication, the resistance / tolerance and cycle of the cultivar used (ALMEIDA, 2005).

With the reduction in the efficiency of fungicides registered for the control of Asian rust *(Phakopsora pachyrhizi),* the search for soya genotypes with greater tolerance/resistance to pathogen attack has been an important tool for genetic improvement, with the aim of reducing the impact this disease has on soya crop yields (MARQUES et al., 2014).

INÓX® technology is relatively new and one of its main features is resistance to Asian soya rust (ASR). Launched by the MT Foundation in 2008 in the cerrado region (HIROMOTO & CAMACHO, 2008), it was only in the 2011/2012 harvest that genotypes with this technology began to be cultivated in southern Brazil.

The genotype x environment (GxE) interaction is defined as the differential behaviour of genotypes as a function of environmental diversity. Within this context, the soil and climate conditions associated with cultural practices, the occurrence of pathogens and other variables that affect plant development characterise the environment. From this, it can be considered that the environment is made up of all the factors that affect plant development that are not of genetic origin (BORÉM & MIRANDA, 2009).

The aim of this study was to assess the performance of soya resistant to Asian rust in different environments in the state of Rio Grande do Sul. The experiment was conducted in the 2013/2014 agricultural year in five soybean-growing regions in the state of Rio Grande do Sul.

The environments where the experiment was carried out were located in the state of Rio Grande do Sul: Independência, located under the coordinates 27°51'18"S and 54°17'13"W, altitude 315m and soil type Latossolo Vermelho distrófico; Tapera, located under the coordinates 28°42'H"S and 52°51'25"W, altitude 381m and soil type Latossolo Vermelho distroférrico; Derrubadas, located under the coordinates 27°16'63"S and 53°47'33"W, altitude 430m and soil typeNitossolo Vermelho Eutroférrico argissólico; Frederico Westphalen, located under the coordinates 27°39'05"S and 53°42'94"W, altitude 490m and soil type Latossolo Vermelho distrófico and Pelotas, located under the coordinates 31°52'00"S and 52°21'24"W, altitude 13m and soil typeArgissolo Vermelho Amarelo distrófico. The characterisation of the environments followed the soil classification proposed by EMBRAPA (2006), and the climate classification *is* cfa, as proposed by Kõppen.

The experiment was conducted in a randomised block design with four replications. The environments tested were the municipalities of Frederico Westphalen, Derrubadas, Tapera, Independência and Pelotas.

The experiment used the TMG 7161 RR soya cultivar, which has INOX® technology and is characterised by its indeterminate growth habit and physiological maturity group 5.9. The sowing density used was 250,000 seeds per hectare.

The experimental units consisted of five three-metre rows, with 0.45 metre spacing between rows, giving a total plot area of 6.75 m^2 . The useful area of the plot was made up of the three centre rows, discarding 0.5 m from each end, giving a total useful area of 2.70 m^2 .

The experiment was carried out in areas previously cultivated with wheat during the winter period. The areas were desiccated beforehand and the furrows were made using no-till seed drills. At the same time, 300 kg ha^{-1} of NPK 02-25-25 was fertilised. Sowing was carried out manually, with the seeds placed three to five centimetres deep. The sowing density used was that recommended by the breeders for each cultivar, respecting the sowing season and the soybean macro-region. The trials were sown between 15 and 19 November 2013.

For the purposes of the experiment, all the environments and experimental units received the same cultural treatments, adopting preventive management to control weeds, pests and diseases. Control was then carried out using products recommended for the crop.

The experiment was harvested between 25 March and 5 April 2014. The experimental units were harvested by hand and the grains were threshed using an electric grain thresher.

The variables of agronomic interest assessed on the basis of the useful area of the plot were: Grain yield - RG, the result extrapolated to kg ha^{-1} , from the total grain mass of each experimental unit, corrected to 13% moisture. Thousand-grain mass - MMG, measured by manually counting eight replicates of 100 seeds, expanded to thousand-grain mass, results in grams (g). The yield components were assessed by taking 10 representative plants from the useful area of the plot: Grain mass per plant - MGP, measured by the mass of grains per plant, results in grams (g). Number of legumes on the stem - NLH, measure of the number of legumes on the stem per plant. Plant height - AP, measured from the base of the plant with the ground to the end of the main stem of the plant, results in centimetres (cm). Height of insertion of first legume - AIPL, measured from the base of the plant with the soil to the first legume on the stem, results in cm. Number of Legumes on Branches - NLR, a measure of the total number of vegetables present on the branches per plant. Number of branches - NR, measured by counting the total number of branches longer than 10 centimetres present on a plant. Number of vegetables with one grain, obtained by counting all the vegetables with one grain on a plant. Number of vegetables with two grains, obtained by counting all the vegetables with two grains present on a plant. Number of vegetables with three grains, obtained by counting all the vegetables with three grains present on a plant.

The analysis of variance revealed significance by the F test for the variables grain yield, mass of grains per plant, mass of a thousand grains, plant height, height of insertion of the first legume, number of branches, number of legumes on the branch, number of legumes with two grains, number of legumes with three grains. No significant differences were found for the variables number of legumes on the main stem and number of legumes with one grain.

Among the five evaluation environments, Derrubadas had the highest grain yield, while Pelotas had the lowest magnitudes for this variable. From this, it can be seen that although the TMG 7161 RR genotype has agricultural zoning for all the environments under study (BRASIL, 2014), its behaviour differed between them.

Borém & Miranda (2009) attribute this behaviour to the genotypes x environment interaction, i.e. the differential behaviour shown by genotypes as a function of environmental diversity, which is influenced by soil and climate conditions associated with cultural practices, the presence of pathogens and other factors that affect plant development. According to Rezende & Carvalho (2007), environmental factors such as humidity, temperature and photoperiod determine the genotype used in a region and have a direct influence on the expression of the plant's maximum genetic potential.

Similarly, for the grain mass per plant variable, higher magnitudes were found in Derrubadas. It can be inferred that MGP had a positive correlation with grain yield, corroborating Dalchiavon & Carvalho (2012), who showed that grain mass per plant and the number of legumes per plant are directly associated with soya yield, so that these traits can be used successfully to estimate soya yield.

For the thousand grains variable, better results were found in Pelotas. This yield component is strongly affected by the availability of water during the grain development and filling phase. According to Pandey & Torrie (1973), average grain weight is genetically determined but influenced by the environment. Corroborating this, in studies applying water restriction to soya plants during the grain filling phase, Sionit & Kramer (1977) observed a decrease in the weight of a hundred grains. When water restriction occurs around a month after flowering, studies show that there is a reduction in grain size, and if it occurs during flowering, there is a reduction in the number of legumes, thus increasing grain size (MOMEN et al., 1979). In studies carried out by Egli et al. (1983), there was an increase in the mass of a hundred grains when water restriction was applied before the R5 reproductive stage, and a reduction when this stress was applied from the R5 stage onwards.

For the plant height variable, the highest magnitudes were found in Frederico Westphalen, with 119.63 cm, while the lowest results were found in Independência, with 81.10 cm. According to Garcia et al. (2007), it is desirable for the final plant height to be above 60 cm, as this helps to reduce grain loss during harvesting. It can be inferred that, although there were differences between the environments, in all of them the genotype under study had a plant height above the minimum level mentioned by this author. However, Souza et al. (2013) point out that more compact, more balanced plants can be more efficient at photosynthesis, while soya research has been looking for smaller plants with a more balanced architecture that are able to support the large number of pods and grains until harvest time (SINGH, 2001).

The highest height of insertion of the first legume among the environments was found in Tapera. On the other hand, the lowest height among the five environments studied was found in Pelotas. However, all environments had a first legume insertion height within the range recommended by EMBRAPA (1996), which states that a first legume insertion height greater than 12 cm minimises harvest losses.

No significant differences were found in the number of legumes on the main stem for the TMG 7161 RR genotype in the five different environments under study. This corroborates Bahry et al. (2011), who, when studying nitrogen fertilisation in the reproductive stages on the agronomic development of soya, found no significant difference between the dose factor and times of nitrogen application on the number of legumes on the main stem.

When analysing the variable number of branches (NR), it can be seen that the highest magnitudes were obtained in Derrubadas, reaching 3.17 branches per plant. Navarro Júnior & Costa (2002) point out that the morphological characteristics number of branches per plant, length of branches and number of fertile nodes are directly related to the plant's productive potential, as they represent a greater photosynthetically active area and also potentially productive through the number of environments for the development of floral primordia. From this, it can be inferred that the greater number of branches achieved in Derrubadas contributed to a higher grain yield in this environment.

The lowest number of branches per plant was observed in Frederico Westphalen, while the highest plant height values were found in the same environment. This corroborates Mauad et al. (2010), who, in studies on soya bean sowing density, attributed the variations in the number of soya bean branches to the competition that occurs between soya bean plants for the growth factors in the environment, especially for light, where at higher sowing densities, due to the excessive number of plants in the row, there is less availability of photoassimilates for the growth of the branches, which are preferentially destined for the growth of the main stem.

For the variable number of legumes on the branches, the Independência, Derrubadas and Pelotas environments were superior to the other environments, while they did not differ from each other. Similarly, Gubiani (2005) assessed the response of soya beans to sowing times and plant arrangements and found differences in the number of legumes on the branches.

With regard to the number of legumes per plant with 1, 2 and 3 grains, it can be seen that Derrubadas obtained a higher number of grains with 2 and 3 grains, reaching magnitudes of 23.52 and 25.97 legumes per plant, respectively. However, no significant differences were found for the number of legumes with 1 grain. This response may be related to the environmental and climatic conditions in the Derrubadas environment, which enabled the TMG 7161 RR genotype to express the number of grains per legume in greater magnitude, given that Heiffing (2002) considers the number of grains per pod to be a typically genetic characteristic. Mcblain & Hume (1981) emphasise that the number of grains per legume is strongly influenced by the fact that modern cultivars are selected for the formation of 3 ovules per legume.

Soya resistant to Asian rust does not perform equally well in all the growing environments, possibly because the environments have contrasting soil and climate conditions and even a resistant cultivar fluctuates in terms of the parameters assessed, indicating that more restrictive agricultural zoning should be carried out.

The performance of soya resistant to Asian rust *(Phakopsora pachyrhizi) is* best expressed in Derrubadas under the variables grain yield, grain mass per plant, number of branches and number of legumes with two and three grains.

CHAPTER 2

Adaptability of soya grown on Planossolo in the far south of Brazil

Ivan Ricardo Carvalho, Vinícius Jardel Szareski, Maicon Nardino, Velei

Queiroz de Souza

Soya is one of the world's main oilseed crops, used as a raw material for food and feed. In the 2014/2015 agricultural harvest, around 31.3 million hectares were sown in Brazil, producing 93 million tonnes of grain (CONAB, 2015). In Rio Grande do Sul, 4.9 million hectares were cultivated, of which 302,000 hectares are in lowland areas in succession to rice (IRGA, 2014).

Soybean genetic improvement programmes are currently aimed at increasing productivity while directly and indirectly reducing production costs. At the same time, the aim is to increase the adaptive capacity of genotypes in relation to soil and climate conditions and the incidence of insect pests and diseases (NASCIMENTO et al., 2010; GULLUOGLU et al., 2011).

Among the factors that influence the growth and development of soya, photoperiod directly influences the crop's productive potential (VERNETTI, 1983). Research by Sanchez (2012) shows that changes in soya morphology can be due to altitude, latitude, soil texture and fertility, plant population, sowing time, spatial arrangement between rows and intrinsic characteristics of the genotype.Rezende and Carvalho (2007) indicate that among these factors, sowing time is closely related to the response to photoperiod. As such, when soybean genotypes are introduced to a new region, they must undergo preliminary analyses of the changes in the growing environment and their responses to environmental variations (CRUZ & REGAZZI, 2001).

In order to understand the genetic variation between the genotypes grown in a given environment, canonical variables can be used to provide a simplified and joint interpretation of the characters measured, which is presented through a graphical dispersion of the two- or three-dimensional scores. This dispersion allows groups of similar genotypes to be visualised and genetically dissimilar groups to be discriminated (CRUZ et al., 2014).

There is a lack of information on the performance of soya genotypes grown in the lowlands of Rio Grande do Sul, so the aim of this work was to make inferences about production performance, the linear associations of characters and to identify the genetic variation of soya genotypes using canonical variables.The experiment was conducted during the 2013-2014 agricultural season in the municipality of Capão do Leão - RS, with geographical coordinates of 31°52'00"S and 52º 21'24"W, and an altitude of 13.24m. The climate, according to Kõppen, is subtropical *Cfa*. The soil is classified as Argissolo Vermelho Amarelo Distrófico, with a sandy loam texture (SANTOS et al., 2006).

The experimental design was randomised blocks with four replications. The genotypes used were: NS 5445 IPRO, NS 6211 RR, TEC 6029 IPRO, TECS 13/03 RR, 6458 RSF IPRO, DON MARIO 5.8i, 6160RSF IPRO, DON MARIO 5.9i, CD 2611 IPRO, CD 2585 RR and TMG 7161 RR.The experimental units were made up of five three-metre rows and 0.45-metre spacing between rows, giving a total area of 6.75 m^2 . The useful area considered was the three centre rows, discarding 0.5 m from each end and 2.70 m^2 .

Sowing was based on the no-till system, with a base fertiliser of 300 kg ha^4 of NPK in the 02-25-25 formulation. The population density used was 300,000 plants per hectare. Control of invasive plants, insect pests and diseases was carried out according to the crop's needs. The yield components were assessed by randomly sampling 10 representative plants from each experimental unit, and the characters were measured using the methodology proposed by Carvalho et al. (2015): Number of legumes on the main stem (NLH), results in units; Plant height (AP), results in centimetres (cm); Insertion height of the first legume (AIPL), results in cm; Number of legumes on the branches (NLR), results in units; Number of branches (NR), results in units; Number of legumes with one grain (LUM), results in units; Number of legumes with two grains

(LDOIS), results in units; Number of legumes with three grains (LTRE), results in units; Grain mass in ten plants (MlOP), results in grams (g); Thousand grain mass (MMG), results in grams (g); Grain yield (RG), results in kg ha[1] .

The analysis of variance revealed significance for the characters thousand grains mass (MMG), grain yield (RG), grain mass in ten plants (M10P), plant height (AP), number of legumes with one grain (LUM), number of legumes with two grains (LDOIS) and number of legumes with three grains (LTRE). The characters that did not differ statistically between the genotypes tested were: number of legumes on the main stem (NLH), height of insertion of the first legume (AIPL), number of legumes on the branches (NLR) and number of branches (NR).

The thousand grain mass (MMG) was higher for the TEC 6029 IPRO genotype, which differed from the ND 6211 RR, TEC S 13/03 RR, CD 2585 RR and TMG 7161 RR genotypes. The same genotype was 33% superior to the TEC S 13/03 RR genotype, which had the lowest thousand-grain mass. Work carried out by Navarro Júnior and Costa (2002) found that the genotype with the highest thousand grain mass was 22.5% superior to the genotype with the lowest, in studies involving six soya cultivars in the municipality of Eldorado do Sul - RS. Brandt et al. (2006) found that the soya genotype with the highest and lowest magnitudes was 14.6% superior to the genotype with the lowest magnitude, working with nine crop succession systems in Dourados - MS, indicating that this character shows some variation.

The mass of a thousand grains is genetically governed by many genes, but is highly influenced by the environment (NAVARRO JÚNIOR & COSTA, 2002). Rambo et al. (2004) observed that the mass and size of the grains increased when the spacing between plants became equidistant, together with a reduction in plant population density, which directly affects the spatial arrangement and intra-specific competition of the soya bean. Water deficits during the grain filling period affect the mass of these grains, reduce photosynthetic capacity, cool tissues and impair the accumulation of photoassimilates, consequently jeopardising their transport to the grains (FONTOURA et al., 2006). Another factor that influences this character is the earliness of the genotypes, due to low latitude regions altering the photoperiod, providing less vegetative development and also less distribution of photoassimilates, resulting in lighter grains with smaller dimensions (MAEHLER et al., 2003).

Grain yield (GY) was highest for the 6160RSF IPRO genotype, but only differed from the Don Mario 5.9i genotype. This is a character that depends on the interaction of various factors, including soil characteristics, mineral nutrition, climatic conditions, altitude, latitude, photoperiod and all the management used from sowing to harvest. Currently, one of the main objectives in soybean breeding programmes *is to* find superior genotypes with high yields. However, this *is* a quantitative trait, controlled by several genes and greatly influenced by environmental conditions (REIS et al., 2001). Medeiros et al. (1991) stated that possible causes of low grain yields in soya beans are attributed to the soil and climate characteristics of the region and the level of technology used by the producer.

The character mass of ten plants (M10P) has a similar meaning to grain yield (RG), where the 6160RSF IPRO genotype showed superiority, but differed only from the NS 6211, Don Mario 5.9i, CD 2611 IPRO and CD 2585 RR genotypes, so interpreting this character together with grain yield (RG) makes it possible to infer greater reliability in the response of the genotypes, since even by sampling plants in the experimental unit the results match those obtained by harvesting the entire useful plot.

The plant height (PH) shows that the genotypes behave similarly, with the exception of the NS 6211 RR genotype, which had a 34.4% lower height than the genotype with the highest magnitude. The values ranged from 65 to 102.13 cm, which is considered ideal for mechanised harvesting, and the results are in line with Rezende and Carvalho (2007) who state that in order to reduce losses during mechanised harvesting, soya beans should be between 60 and 120 cm tall.

Changes in the growth habit of soya beans are evident in the new genotypes, where the determinate growth habit is less favoured by producers when compared to genotypes with indeterminate growth habit, as these have a longer juvenile period and a taller size, which makes early sowings possible and provides an escape from certain periods of water stress that occur frequently in the southern region (EMBRAPA,

2008).Plant height depends on the elongation of the internodes and the number of nodes on the stem (TAIZ & ZEIGER, 2004). Guimarães et al. (2008) state that even though genotypes are insensitive to photoperiod and do not change their flowering, they do show changes in plant height.

In relation to the number of legumes containing one (LUM) and two (LDOIS) grains, the 6160RSF IPRO genotype had higher magnitudes, which differed from the CD 2611 IPRO genotype. For the character number of legumes containing three grains (LTRE), the greatest magnitude was found in the genotype 6458RSF IPRO, which differed from the genotypes NS 5445 IPRO, NS 6211 RR, Don Mario 5.8i and CD 2585 RR.Results by Navarro and Costa (2002) identified the contribution of the number of legumes and grains per leg to soya bean yield in Eldorado do Sul - RS. According to Rambo et al. (2004), the three strata of the soya bean canopy are modified as a result of the change in plant arrangement and water demand, where the number of grains per legume (one, two and three grains) are the most influenced characters.

Considerable influences on the number of grains per legume are attributed to inter- and intra-specific competition and water deficits, which modify the balance between the production of flowers per plant and the proportion of these that develop and form legumes (NAVARRO & COSTA, 2002). According to Mundstock & Thomas (2005), the number of grains per legume shows little phenotypic variation, demonstrating the uniformity of genetic improvement in the search for plants with two and three grains per legume. The most productive genotype in this study had the highest number of legumes with two grains, thus corroborating the results found by Pípolo et al. (2005).

Pearson's linear correlation shows a trend between the characters measured, and its magnitudes range from -1 to 1, with negative and positive coefficients, so the closer to 1 or -1 the stronger the degree of linear association between the two variables and the closer to zero the lower the linear association (CRUZ et al., 2012). According to Carvalho et al. (2004), the coefficients are interpreted as follows: r= 0 indicates no correlation.Plion; r= 0 to 0.3 indicates a weak or low relationship; r= 0.3 to r= 0.6 indicates a medium or intermediate correlation; r=0.6 to r=1.0 indicates a strong or high relationship. The same interpretations apply to negative associations.

Pearson's linear correlation was carried out between 11 characters, with 55 associations, 32 of which were significant pairs. Grain yield (RG) showed a high and positive linear correlation with M10P (r=0.93), and a medium and positive correlation with the characters NLH (r=0.44), NLR (r=0.39), LUG (r=0.43), LDG (r=0.40) and LTG (r=0.47), indicating that the greater their magnitude, the greater the increase in grain yield. The same applies to the character mass of ten plants (M10P), which showed an average positive linear correlation with AP (r=0.31), NLH (r=0.41), NLR (r=0.50), NR (r=0.35), LUG (r=0.47), LDG (r=0.50) and LTG (r=0.53).

Plant height (PH) showed a strong positive correlation with the characters AIPL (r=0.65) and NLH (r=0.65), and a medium positive correlation with the character number of legumes with three grains LTG (r=0.34), which means that the higher the PH, the higher the height at which the first legume is inserted and the greater the number of legumes on the main stem.

For the character height of insertion of the first legume (AIPL), there was an average positive correlation with the character NLH (r=0.36). On the other hand, there was an average negative correlation with NLR (r=-0.36), NR (r=-0.40) and LDG (r=-0.41). Studies by Mauad et al. (2010) show that soya beans have the capacity to change morphologically according to the growing environment. This is a very important feature when it comes to harvesting, as plants with pulses very close to ground level incur losses due to the height of the cutting bar not allowing the plant's pulses to be fully harvested. According to Sediayama et al. (1999), the minimum height of the first vegetable should be 10 to 12 cm in soils with flat topography and 15 cm in soils with steeper slopes, in order to minimise losses.

The number of legumes on the main stem (NLH) showed a positive and medium correlation with LUG (r=0.33), LDG (r=0.30) and a weak and positive correlation with LTG (r=0.29), which showed that increasing the magnitude of this trait could possibly result in a higher number of grains per plant. The number of legumes on the branches (NLR) showed a strong positive correlation with NR (r=0.87), LUG (r=0.60), LDG (r=0.82) and LTG (r=0.64). Likewise, the number of branches (NR) character was positively and strongly correlated with LDG (r=0.64) and moderately correlated with LUG (r=0.38) and LTG (r=0.49). This indicates that an increase in the number of branches may lead to a greater number of legumes per plant. Several studies have shown that the number of legumes is one of the most dynamic components, which can be the key to increasing productivity, especially when there is a change in sowing density and a reduction in row spacing (PEIXOTO et al., 2000). This character, together with the length of the branches, is related to yield potential, since they represent a larger photosynthesising area, with more flowers and vegetables. On the other hand, their length may divert photoassimilates and represent an additional demand for fixing the reproductive structures in the plants (NAVARRO & COSTA 2002).The number of legumes with one grain (LUG) correlated positively and strongly with LDG (r=0.64), while the number of legumes with two grains (LDG) showed a positive and medium correlation with LTG (r=0.33).

Using canonical variable analysis, *it is* possible to group genotypes based on characters of agronomic interest (CRUZ et al., 2012). The analysis of canonical variables was carried out on the basis of the traits thousand grain mass, grain yield, mass of ten plants, plant height and number of legumes with one, two and three grains determined in 11 soya genotypes grown in the 2013/2014 agricultural season. Variation in the genotypes was expressed through the first two canonical variables, where *VC1* accounted for 68.70% and *VC2* 16.31% of the total variation.

Four groups were formed, where the first group was made up of genotypes 1 (NS 5445 IPRO) and 2 (NS 6211 RR), the second group was made up of genotype 9 (CD 2611 IPRO) which has an early cycle, indeterminate growth habit and belongs to maturity group 6.1; the third group was made up of genotype 10 (CD 2585 RR) which has an indeterminate growth habit, a super-early cycle, is suitable for high and cold regions and belongs to maturity group 5.The fourth group was made up of genotypes 3 (TEC 6029 IPRO), 4 (TEC S 13/03 RR), 5 (6458RSF IPRO), 6 (6160RSF IPRO), 7 (DON MARIO 5.8i), 8 (DON MARIO 5.9i) and 11 (TMG 7161 RR).

The canonical variables obtained explain 85.01% of the variation involved in the study, so reliable interpretation by this multivariate analysis must be based on minimum estimates of 80% across the set of characters (CRUZ et al., 2006). Therefore, in order to identify which genotypes are superior in a competition trial, it is feasible to use univariate and multivariate analyses together, which results in adequate answers as to which genotypes are superior, genetically closer, and determine the characters of interest in terms of grain yield. This genetic variability obtained between genotypes is essential for breeders, as it makes it possible to effectively select which genotypes are most suitable for certain crops (BERNARDO, 2002).

The 6160RSF IPRO, NS 5445 IPRO and 6458RSF IPRO genotypes have the highest magnitudes for the grain yield character, showing potential for use in lowland crops in Rio Grande do Sul. The characters mass of ten plants, number of legumes on the stem, number of legumes on the branches, number of legumes with one, two and three grains have positive linear correlations with grain yield. The genotypes are grouped according to their genetic basis and ripening cycle, where the canonical variables explain 85.05% of the existing genetic variation, and form four groups of genotypes.

CHAPTER 3

Soya seed yield components depending on growth habit

Maicon Nardino, Ivan Ricardo Carvalho, Vinícius Jardel Szareski,
Francisco Amaral Villela

Soya *(Glycine max* L.) is characterised as a legume belonging to the Fabaceae family and comes from Asia. It was introduced to Brazil in the 1970s and is currently the main commodity grown. This oilseed has high nutritional value and grain production, and can be used as green manure, fodder, silage and hay (Sediyama, 2009). According to Conab (2014) soya in the 2013/2014 crop year shows an 8.7% increase in cultivated area and grain yields of around 2,860 kilos per hectare.

The yield potential of soya beans is determined by the contribution of various agronomic characters, such as the number of plants per unit area, legumes per plant, grains per legume, grain mass, branches per plant, reproductive nodes and branch length (Thomas and Costa, 2010). It is therefore necessary to know the relationships between these traits, defining their contribution to grain yield.

Associations between characters are established using the correlation coefficients of the path analysis (Wright, 1921), which provides an understanding of cause and effect relationships, both directly and indirectly on the main variable. According to Ramalho et al. (1993), correlation coefficients make it possible to identify the changes of a given character on another. Defining the interrelationships between these traits can contribute to genetic improvement through the indirect selection of traits with less complex inheritance that promote gains in soya production potential.

The classification of soya genotypes in terms of genetic distance and kinship can be gauged through genetic divergence based on the individual's phenotype, which can quantify the similarity and dissimilarity between genotypes, making it possible to distinguish between groups and the extent to which these groups contrast in their characteristics. Distinguishing between groups can contribute to the choice of genitors and increase genetic variability in the segregating population (Cruz et al., 2004). The aim of this study was therefore to determine the direct and indirect phenotypic effects of characters of agronomic interest linked to the grain yield of soya beans with indeterminate growth habit and the genetic dissimilarity between the genotypes. The experiment was conducted in 2013/2014 at the Federal University of Santa Maria, Frederico Westphalen *campus*, in the Laboratory of Genetic Improvement and Plant Production, located at coordinates 27°39'S, 53°42'O, with an altitude of 490 metres. According to Kõppen, the climate is characterised as humid subtropical Cfa, and according to Mota (1953) the soil is classified as Ferric Aluminosilicate Latosol.

The experimental design used was randomised blocks, with ten soya cultivars with indeterminate growth habit, arranged in three replications. The cultivars used were FPS Paranapanema RR, BMX Classe RR, FPS Solimões RR, BMX Potência RR, BMX Força RR, BMX Energia RR, BMX Turbo RR, FPS Iguaçu RR, BMX Tornado RR, BMX Alvo RR. The experimental units consisted of four rows three metres long, spaced 0.45 metres apart, with a population density of 300,000.00 plants ha^1 . The crop was planted in a direct sowing system, using a base fertiliser of 200 kg ha^1 of N-P-K in the 5-20-20 formulation. Pests and diseases were controlled preventatively. Evaluations were carried out on the centre lines of each experimental unit, disregarding the first metre at each end in order to reduce the effects of the border. Ten plants were collected at random to measure the characters of agronomic interest, making up the average of each variable in the experimental unit. The characters assessed were:

- Insertion of the first legume (IPL): measured from the neck of the plant to the insertion of the first legume, in centimetres.

- Plant height (PH): measured from the neck to the apex of the plant, in centimetres.

- Number of vegetables on the main stem (NLHP): total number of viable vegetables on the main stem.

- Number of vegetables on the branches (NLR): total number of vegetables on the branches.

- Number of vegetables per plant (NLP): sum of the vegetables on the main stem and branches.

- Number of reproductive nodes on the main stem (NNHP): count of nodes that have produced viable vegetables on the main stem.

- Number of reproductive nodes on the branches (NNR): count of nodes that have emitted viable vegetables on the branches.

- Number of branches (NR): total number of branches per plant.

- Branch length (BR): measured from the insertion of the branch to the apex, considering those longer than ten centimetres.

- Number of vegetables per node on the main stem (NLNH): ratio between the total number of vegetables on the main stem and the number of reproductive nodes.

- Number of legumes per node on the branches (NLNR): ratio between the total number of legumes on the branches and the number of reproductive nodes.

- Intemode length (CINT): ratio between the number of nodes on the main stem and the length of the branches, in centimetres.

- Thousand-grain mass (GGM): measured by sampling eight sub-samples of 100 grains per experimental unit, after which the character was averaged in grams.

- Grain yield (REND): measured using the total mass of grains in the experimental unit, with subsequent correction of the humidity to 13%, the grain mass was adjusted using the number of plants per experimental unit, obtaining the corrected grain yield per plant, after which it was adjusted for the plant population used, in Kg ha[-1] .

Grain yield is characterised as a quantitative trait controlled by the joint action of several genes, which contribute in small proportions to the performance of the phenotype, but are directly dependent on the environment (Allard, 1971). Therefore, genetic improvement of soya in order to obtain superior genotypes must be based on knowledge of the characters and their interconnections. These relationships show the direct and indirect effects of characters on grain yield (Coimbra et al., 1999).

Multicollinearity can express high coefficients, resulting in values that are unrelated to the biological phenomenon being studied (Coimbra et al., 2005). In this case, variables are eliminated and the methodology proposed by Carvalho (1995) of least squares or Christian path analysis is applied. When assessing the multicollinearity of the phenotypic correlation matrix, severe multicollinearity effects were observed for the following characters: number of nodes on the main stem, number of nodes on the branches, number of vegetables per node on the main stem and number of vegetables per node on the branches. The characters were therefore removed from the analysis in order to avoid the effects of multicollinearity. The number of vegetables per node on the main stem was included.

Soya genotypes with indeterminate growth habit show direct influences on grain yield through the characters thousand grains, branch length, number of legumes on the main stem, plant height and intemode length. Studies carried out by Nogueira et al. (2012) revealed moderate to high correlation coefficients for grain yield with direct effects through the characters number of legumes per plant, mass of one hundred grains and number of grains per legume. According to Kurek (2001), in beans the direct contributions to grain yield are revealed through the number of legumes per plant and grain mass.

The insertion of the first legume influences the mechanised harvesting process, where lower magnitudes of this character lead to more harvest losses (Braz et al., 2010). The direct effects showed low and negative correlation coefficients, revealing that genotypes with a lower height of insertion of the first legume tend to increase grain yield through the number of reproductive nodes or branches on the main stem. In terms of indirect effects, there were low and positive correlation coefficients for plant height, and low and negative correlation coefficients for the number of legumes on the main stem and intemode length. Therefore, increases in the height of insertion of the first legume allow the genotype to increase its height, but do not influence grain production. Almeida et al. (2010) revealed positive effects between the height of insertion of the first legume and grain yield. Pearson's correlation shows intermediate and negative coefficients (r= - 0.30).

The plant height shows direct effects, with a low and positive correlation coefficient in relation to the main variable. A low and positive indirect effect *is* evidenced by the length of the intemode. Therefore, an increase in soya bean height leads to elongation of the main stem and distancing between the reproductive nodes. Low and negative indirect effects are observed through the insertion height of the first legume. Studies by Coimbra et al. (2004) show no relationship between plant height, thousand grain mass and grain yield. According to Souza et al. (2013), lower statures in soya beans result in an increase in the number of grains per legume, grains per plant and grain mass, contributing to the crop's productive potential. Pearson's total correlation (r=0.219) was low and positive, proving the relationship between the explanatory variables and the main parameter.

The number of legumes per plant makes a major contribution to the productive potential of soya beans, and fluctuates greatly depending on the growing environment, with interventions from abiotic factors, water availability and photoperiod (Lima et al., 2009). The direct effects have low and positive correlation coefficients with grain yield. Indirectly, a low and positive correlation coefficient is observed through the length of the branches, so the number of vegetables on the main stem combined with the length of the branches makes it possible to increase the total number of vegetables per plant. Low and negative coefficients are revealed indirectly through the length of the intemode and the mass of a thousand soya beans.

The grain yield can be explained by the increase in legumes on the main stem, with a reduction in the distance between reproductive nodes and an increase in branching. Results confirmed by studies by Alcantara Neto et al. (2011) show a high contribution of the number of legumes per plant to grain yield. Pearson's total correlation shows a low positive correlation (r=0.139) between the number of legumes per plant and grain yield.

The number of legumes on the branches shows zero correlation coefficients for the direct effects. For indirect effects, intermediate and positive correlation coefficients are observed for branch length, and low and negative correlation coefficients for number of branches and thousand grain mass. Therefore, soya genotypes with indeterminate habit advocate the formation of fewer branches, which in turn contributes to the increase in total legumes per plant. Pearson's correlation showed a low and negative magnitude (r=-0.221) between the number of legumes on the branches and the main character.

The direct effects of the number of branches per plant show a low and negative correlation coefficient with grain yield. Indirectly, there are low and positive coefficients through the characters plant height and branch length, and low and negative coefficients through first leg insertion and thousand grain mass. In this way, indeterminate soya advocates taller plants with fewer branches, but in return reduces grain mass. According to Barbosa et al. (2014), soya beans show the ability to change in relation to environmental conditions, nutritional management and sowing time, mainly altering the magnitude of the characters number of reproductive nodes on the main stem and branches per plant. Pearson's correlation is intermediate and positive (r=0.359), proving the relationship between the magnitude of branches and the productive potential of soya.

The direct effects of branch length show intermediate and positive correlation coefficients with grain yield. The indirect effects show low and positive correlation coefficients via the number of legumes on the main stem, and low and negative correlation coefficients via the mass of a thousand grains. Pearson's correlation shows an intermediate and positive correlation (r=0.405).

Intemode length shows direct effects with low and positive correlation coefficients with grain yield. Indirectly, intermediate to low coefficients are observed, but both are positive through the mass of a thousand grains and plant height. Low and negative indirect effects are observed through the insertion of the first legume, the number of legumes on the main stem and the length of the branches. A study by Souza et al. (2013) revealed high and positive correlation coefficients for the characters number of legumes, grains per plant and thousand grain mass in relation to soya bean yield. Pearson's correlation was low and positive (r=0.272), justifying the relationship between the length of the intemode and the main character.

The mass of a thousand grains is one of the most important traits in terms of soya production potential. Thus, direct effects, intermediate and positive coefficients are observed for the main variable. Studies show that grain size and mass influence grain yield (Coimbra et al., 1999). Indirect effects include low and positive coefficients for the length of the intemode, low and negative coefficients for the number of legumes on the main stem and the length of the branches.

In this way, heavier grains can be obtained from genotypes with fewer legumes and branches per plant. Thus, the formation of fewer energy-intensive structures can contribute to better partitioning of assimilates between grains, favouring an increase in the size and mass of these grains and consequently contributing to soybean yields. A study by Souza et al. (2014) points out that the potential of soybeans is due to leaf expansion, interception of solar radiation, assimilation, conversion efficiency and transport of photoassimilates to reproductive structures. Pearson's correlation (r=0.539) was intermediate and positive in relation to the mass of a thousand grains and grain yield.

The phenotypic estimates of cause and effect in the path analysis proved to be reliable due to the high value obtained for the coefficient of determination (0.859) and low residual effects of (0.376). This highlights the importance of identifying the direct and indirect effects between the explanatory characters and the main parameter in soya beans with indeterminate growth habit.

The purpose of the Singh method is to determine the relative contribution of the agronomic characters that most influence genotype differentiation. Thus, the dissimilarity of the 10 soya genotypes is based on the characters number of legumes per node on the branches with a 57.50% contribution, followed by number of branches with 12.30% and number of nodes per branch 11.66%. In contrast, studies by Almeida et al. (2011) revealed that the variable mass of one hundred grains contributed 26.56% to the genetic dissimilarity between soya genotypes.

The characters plant height, number of vegetables on the main stem, number of nodes on the main stem, branch length, number of vegetables per node on the main stem, number of vegetables per plant and thousand-grain mass all contributed magnitudes of less than 1.00% to genotype discrimination. In studies carried out by Almeida et al. (2011) showed that the number of vegetables per plant contributed less to genetic dissimilarity.

The purpose of cluster analysis is to form groups of genotypes that contrast in terms of their characteristics. In this way, genetic dissimilarity is estimated using the generalised Mahalanobis distance, which makes it possible to differentiate the genotypes into three groups. Group I is characterised as the largest group, comprising six genotypes, such as FPS Paranapanema RR, BMX Potência RR, FPS Iguaçu RR, BMX Turbo RR , BMX Energia RR , BMX Classe RR. Group II is made up of three genotypes: BMX Alvo RR, BMX Tomado RR and BMX Força RR. Group III is made up of just one genotype: FPS Solimões RR, which is the most distant from the other genotypes studied.

According to Bueno (2006), knowledge of the dissimilarity between genotypes can contribute to the choice of genitors that will make up a crossing block in a genetic improvement programme. This is of paramount importance to the success of the activity as it allows combinations between genitors to be directed, where hybridisation between more dissimilar genitors can lead to an increase in genetic variability in the segregating population. The path analysis revealed that the characters thousand grains mass and branch length had direct effects with a high correlation coefficient with grain yield. Significant indirect effects are revealed through the length of the intemode.

The generalised Mahalanobis distance reveals the formation of three groups, where genotypes belonging to groups I and III are more distant and can increase genetic variability in the formation of a crossing block.

CHAPTER 4

Bean seed yields in north-west Rio Grande do Sul

Vinícius Jardel Szareski, Ivan Ricardo Carvalho, Maicon Nardino, Velei

Queiroz de Souza

The bean crop *(Phaseolus vulgaris) is* suitable as a source of protein and energy in its grains in different regions of the world, and its importance for the human diet can be highlighted (GOMES JÚNIOR et al. 2005). Brazilian bean productivity is far below the potential of the cultivars, with the average grain yield being 1,000 kg ha^{-1} (CONAB 2015). There are several factors that influence the grain yield of the crop, such as the factors of the growing environment, the characteristics of the genotype used, the technologies employed, the cultivation systems (MINGOTTE et al. 2014), nutritional management, mainly nitrogen (DEMARI et al. 2015) and water supply (CARVALHO et al. 2013).

The search for genotypes that are more productive and adapted to the most varied growing conditions is one of the main interests of genetic improvement. In this way, the aim is to understand the genetic variability available to breeders, and thus better determine which genotypes are agronomically suitable and respond to environmental effects (RAMALHO et al. 2012). Phenotypic differentiations are important for determining which available genotypes meet agronomic needs and lead to gains in crop improvement. In this way, the phenotype is defined as a set of biological information resulting from the expression of genetic characteristics and the actions of the environment (RAMALHO et al. 2012).

While the breeder has genetic variability available in the bean improvement programme, it is necessary to understand the existing genetic variation and discriminate the genetic distance between the genotypes under study. Therefore, in order to help choose which genotypes best represent the agronomic ideotype, multivariate analysis techniques are used to discriminate between genotypes. Among these techniques is the Standardised Euclidean Distance, which indicates genetic dissimilarity by forming groups, indicating the closest and most distantly related genotypes (CRUZ et al. 2012).

Cluster analysis techniques aim to group genotypes into various groups according to certain criteria, so that there is homogeneity within the group and heterogeneity between the groups. These methods are used effectively to guide the selection of genitors to make up the crossing blocks (CRUZ et al. 2012). Multivariate methods that summarise the set of original variables into a few components that involve a large part of the variation found are also very useful for analysing genetic dissimilarity, with principal component analysis being a highlight (CRUZ & REGAZZI 1997).

In view of this, the aim of this work was to determine the dissimilarity between Creole bean genotypes using multivariate techniques in order to select individuals to make up the crossing blocks and to determine which of the genotypes evaluated are closest to an agronomic ideotype.The experiment was conducted during the 2014/2015 agricultural season in Tenente Portela - RS, at the coordinates corresponding to latitude 27°22'10.20" S and longitude 53°45'23.00"O, with an altitude of 420 metres. The soil is classified as typical ferric red latosol, and the climate according to Kõppen is characterised as humid subtropical *Cfa*.

The design used was an augmented block design, where the common treatments consisted of seven witnesses: BRS Estilo, IPR Uirapuru, IPR Tangara, BRSMG Realce, IAC Formoso, IPR Imperador and IPR Tiziu, and the regular treatments consisted of 17 genotypes:LMGPPl, LMGPP2, LMGPP3, LMGPP4, LMGPP5, LMGPP6, LMGPP7, LMGPP8, LMGPP9, LMGPP10, LMGPP11, PORTELA16, PORTELA37, PORTELA38, PORTELA39, PORTELA64, PORTELA65, as proposed by Ramalho et al. (2012).

The experimental units consisted of two 1.0 metre rows spaced 0.45 metres apart. The population density used was ten seeds per linear metre (s m^1), based on the direct sowing system, with a base fertiliser of 300 kg ha^1 of N-P-K in the formulation (10-

20-20). For top dressing, 90 kg ha^1 of nitrogen in amide form was applied at vegetative stage V4. Phytosanitary measures to control weeds, insect pests and diseases were carried out preventively in order to minimise interference in the results of the experiment.

The traits were measured on ten plants per treatment: days to flowering (DPF), plant height at flowering (ALF), days to maturity (DPM), plant height at maturity (ALM), height of insertion of the first legume (IPL), number of branches per plant (NRP), number of legumes per plant (NLP), number of grains per legume (NGL), grain mass per legume (MGL), number of grains per plant (NGP), grain mass per plant (MGP), grain length (CMG), grain width (LGG), based on the methodologies proposed by Silva (2005) and Demari et al. (2015).

Descriptive analyses using frequency distributions revealed the formation of different phenotypic classes for the characters measured. For the character days to flowering (DPF), five phenotypic classes were formed, ranging from 25 to 55 days. The 40 and 45 DPF classes made up 74.5 per cent of the genotypes. Plant height at flowering ranged from 18 cm to 66 cm, with most of the genotypes being between 18 and 30 cm. In this sense, the wide range of variation observed for PA may make it possible to obtain segregating populations with high genetic variation for this trait.

For the variable days to maturity, early genotypes with up to 75 days to maturity and late genotypes with up to 115 days to maturity were observed. The data indicated that there is a possibility of selecting superior genotypes that contrast in terms of cycle, making it possible to insert them into specific market niches that need early genotypes to make the production system viable in rotation with other crops.

Plant height at maturity was one of the variables with the greatest variation between the genotypes, ranging from 15 cm to 75 cm AP, which shows the difference between the genotypes studied. The same can be seen for the variable height of insertion of the first legume, with genotypes ranging from 2 cm to 22 cm.

With regard to the number of branches per plant, the largest number of genotypes revealed magnitudes of 0.75 branches per plant.

The variable number of legumes per plant was also contrasting, with genotypes ranging from 3 to 27 legumes being observed, with 75% of the genotypes showing between 3 and 9 legumes per plant. For the number of grains per legume, similar distributions were observed, ranging from 1.3 to 6 grains per legume. The results show the potential for selecting superior genotypes, since the number of legumes per plant (KUREK et al. 2012) and the number of grains per legume (CABRAL et al. 2011) are characters with great potential for indirect selection to increase grain yield.

For the mass of grains per legume, the distribution showed that most of the genotypes had magnitudes of 0.6 to 1.2 g per legume. The number of grains per plant was also highly contrasting, ranging from 1 to 75, which represents the wide variation between the genotypes evaluated. For the mass of grains per plant, it was observed that most of the genotypes were concentrated in magnitudes of 2.5 g to 7.5 g per plant. For grain length, genotypes ranging from 9 to 15 mm were found. For grain width, there were also distributions ranging from 5.7 to 8.1 mm.

The multivariate techniques used to identify the dissimilarity between the genotypes and cultivars revealed that there is dissimilarity between the genotypes studied. According to Table 1, which shows the dissimilarity matrix between the 24 bean genotypes based on Euclidean distance, it can be seen that the greatest distance was 2.25 between the pair of genotypes 19 (PORTELA 16) and 16 (LMGPP09). The highlighted pairs of genotypes represent the greatest distances found, and are indicated as possible genitors to make up the crossing blocks, while genotypes 8 and 9 were the most similar, and or with the smallest distance (0.27). Although this reveals the genetic distance between each combination of genotypes, further analyses are needed to form the crossing blocks in order to identify promising genotypes.

When the genotypes were grouped using the Tocher optimisation method, three distinct groups were formed with the largest inter-group distance of 1.86 ($D_{I, III}$). The largest group formed was group I, made up of 22 genotypes, while groups II and III were made up of one genotype each.

When comparing the results of the Tocher grouping with the genetic dissimilarity matrix, it was observed that the genotypes most distant by the dissimilarity matrix were reliably separated in both multivariate methods. Comparing the constitution of the groups, it can be seen that the cultivars were kept in similar groups, such as the cultivars IPR UIRAPURU, IPR TANGARÁ, IPR IMPERADOR, IPR TIZIU, being group I. The presence of genotypes from the same breeding programme in the same group shows the similarity between them (FALCONER 1987), and possibly only one of the genotypes would be incorporated into the crossing block due to its close relationship with the other genotypes. In addition, the genetic distance (1.34) between cultivars from the same programme that formed different groups in the distance matrix, such as BRS ESTILO (Group I) and BRSMG REALCE (Group II), indicates that genetic constitutions with a greater range of genetic variation can be obtained from the same breeding programme. The formation of groups II (BRSMG REALCE) and III (PORTELA16), made up of just one genotype each, shows their high genetic distance from the other genotypes studied. The distance between group II and III was 1.51, which corresponds to the distance between genotypes 4 and 19.

The significant cofenetic correlation (r) was 0.82. The higher this coefficient, the less distortion there is between the original dissimilarity matrix and the graphical grouping (BUSSAB et al. 1990). In this respect, Sokal & Rohlf (1962) state that coefficients above 0.80 are considered satisfactory, representing reliability of the original matrix. The calculated values of distortion (2.42%) and stress (15.58%) also indicate the reliability of the graphical representation to the original matrix.

As with the Tocher optimisation method, the same genotypes made up the groups. The first group was made up of 22 genotypes, with around 92% of the total genotypes studied, the second by the BRSMG REALCE genotype and the third group by the PORTELA16 genotype. These results show that it is possible to combine these with the genotypes in group I. The most favourable combinations using the commercial witnesses in the crossing blocks would be with genotype 4 (BRMSG REALCE): G4 x G8; G4 x G9; G4 x G16 and G4 x G23, while the most promising combinations using genotype 19 (PORTELA16) would be: G19 x G2; G19 x G3; G19 x G9; G19 x G10; G19 x G16 and G19 x G25. The use of more than one grouping method allowed them to complement each other, generating reliable information from the data and avoiding erroneous inferences that can occur due to the different criteria used by each method.

When assessing the relative contribution of the characters, it was observed that the characters that contributed most to the dissimilarity between the genotypes analysed were height at maturity (31.71%), number of grains per plant (28.47%), days to maturity (16.52%) and height at flowering (12.15%), which explained approximately 90% of the total variation between the genotypes. The characters number of branches per plant, number of grains per legume, mass of grains per legume, length of grain and width of grain did not explain 1% of the total variation found between the genotypes.

With regard to the analysis of the first principal components, the feasibility of using the principal components depends on whether the original set of variables can be summarised in a two- or three-dimensional space without losing information, so the first three principal components were used, which explained approximately 65.56% of the total variation. The three-dimensional projection of the genotypes as a function of the scores estimated for the first three principal components formed groups similar to the Tocher optimisation and UPGMA hierarchical three-group clustering methods. The method made it possible to identify the distance between genotypes 19 and 16 (represented by a triangle), which were also contrasting according to the original dissimilarity matrix.

Although the first three principal components explained only 65.56% of the total variation found, this method proved to be coherent with the other methods for analysing dissimilarity between genotypes, thus complementing the results found in the other methods. In order to exploit the maximum genetic variability in crosses, with a view to obtaining segregating populations, with the possibility of obtaining superior genotypes, the genetic base of the bean shows promise for expansion, considering that the multivariate analysis techniques revealed that there are indications of dissimilarity between the genotypes. In this sense, multivariate methods are promising and useful to breeders in terms of facilitating the interpretation of data, highlighting the dissimilarity of genotypes, and optimising service time as well as human and financial resources.

The genotypes LMGPP09 and PORTELA 16 are the most dissimilar, while LMGPP01 and LMGPP02 are the most similar. The BRSMG REALCE and PORTELA16 cultivars can be used as parents in combinations with other genotypes to form segregating populations with ample genetic variability.

The UPGMA hierarchical clustering method, Tocher optimisation and principal component analysis are coherent and complementary, providing a reliable answer when interpreting data on genetic dissimilarity in beans.

CHAPTER 5

Industrial quality of wheat genotypes in the upper Uruguay region of Rio Grande do Sul

Ivan Ricardo Carvalho, Vinícius Jardel Szareski, Maicon Nardino

Wheat *(Triticum aestivum* L.) is considered one of the main sources of energy for humans, as it provides the necessary fractions of carbohydrates and proteins. This cereal is made up of 55 per cent starch, 12 per cent protein, 1.5 per cent fat and 1.5 per cent mineral material (ROSTAGNO et al., 2011), and is important in production systems as it makes up human and animal feed. Wheat is used as a raw material for biscuits, cakes, pasta and bread (SCHEUER et al., 2011). This is possible because the protein fraction of the grains contains gliadins and glutenins, which when hydrated form gluten, responsible for the elasticity of the dough and fermentation. The quality of food products made from wheat depends on the quality of the flour, protein proportion, gluten strength, extensibility, tenacity, falling number and gluten (ZARDO, 2010).

In this way, the quantity and quality of the protein contained in the grains depends on the soil and climate conditions of the environment and the management techniques used (MAC RITCHIE AND GUPTA, 1993). Research shows that the cultivar, sowing time and density, water management, fertilisers and the interaction between genotypes and environments are determining factors in the quality of wheat flour. Genotypes are more productive and technically demanding, so genetic improvement is seeking quantitatively superior genotypes with high breadmaking quality (WRIGLEY, 1994). Among the resources available to improve the quality of the grains produced, fertiliser management has proven to be efficient, with nitrogen being indispensable for increasing grain yield, which is crucial for protein synthesis and the industrial quality of the grains (SOUSA E LOBATO, 2004).

Among the available nitrogen sources, urea is the main fertiliser used in agriculture; however, it has high losses due to volatilisation of NH_3 , and leaching (CIVARDI et al., 2011; SILVA et al., 2012). On the other hand, ammonium nitrate is an alternative for reducing losses and increasing the efficiency of these nitrogen fertilisers in wheat cultivation (YANO et al., 2005). Given the importance of the technological aspects of wheat grains, and with the aim of increasing quality, the aim of this study was to analyse the technological quality of wheat as a function of the use of different nitrogen sources and managements.

The experiment was conducted in Frederico Westphalen - RS, in the experimental area of the Laboratory of Genetic Improvement and Plant Production at the Federal University of Santa Maria, at the coordinates: 27°23'48.17"S and 53°25'34.82"O, and an altitude of 460 metres. The soil is classified as Latossolo vermelho distrófico (EMBRAPA, 2006), and the climate is characterised by Kõppen as humid subtropical (SEMA, 2005).

The experimental design used was randomised blocks, organised in a factorial scheme: three genotypes (Fundacep 52, Tbio Mestre and Tbio Itaipú) x two nitrogen sources (UR: urea - 45% nitrogen and NA: ammonium nitrate - 33.5% nitrogen) x four managements (AUN: no nitrogen, AFI: tillering, AFR: tillering and flowering and AEF: tillering, rubberisation and flowering), arranged in three replications.

Each plot was made up of 12 rows spaced 0.17 m apart and 3.5 m long, totalling 7.14 m^2 . For the evaluations, eight centre rows were sampled, eliminating 0.5 m from each end. The system was based on direct sowing in the first fortnight of July 2013. The base fertiliser used was 200 kg ha^1 of NPK (08-24-12), and 115 kg ha$^{'1}$ of nitrogen was applied as a top dressing according to the pre-established treatments. The established population density was 310 plants m$^{'2}$. Weeds, insect pests and diseases were managed according to the crop's needs.

The characters were measured at the Cereals Laboratory of the Food Research Centre at the University of Passo Fundo - RS:

Percentage of total protein (PPT): determined using the INFRATEC™ 1241 Grain Analyser, using eight 100 gram sub-samples, results in percentage (%).

Falling number (NQ): determined using the *Falling Number* apparatus, based on method 56-81B of the AACC (2000), using seven grams of sample with 14% humidity, results in seconds (s).

Wet gluten (GU): determined using the Glutomatic apparatus according to method 38-12 of the AACC (1995), using 10 grams of sample, results in percentage (%).

Gluten strength (W): corresponding to the mechanical work required to expand the dough until it breaks, expressed in 10^{-4} Joules.

Tenacity (P): corresponding to the measurement of the maximum overpressure exerted on the expansion of the mass, results in mm.

Extensibility (L): corresponding to the measurement of the length of the curve, predicts the volume of the bread, results mm.

P/L ratio: obtained from the ratio between toughness and extensibility, results in mm mm^{1} .

The analysis of variance revealed significance for the genotype x nitrogen source x management interaction for the characters tenacity (P), extensibility (L), P/L ratio, gluten strength (W), falling number (NQ) and wet gluten (GU). The Tbio Mestre genotype had a higher protein content than Fundacep 52 and Tbio Itaipu. The percentage of total protein was not significantly influenced by the nitrogen sources tested. Protein content cannot be assessed as an indicator of wheat quality in isolation, so extensibility and tenacity are added. Schmidt et al. (2009) revealed that protein content was not an efficient predictor of the industrial quality of wheat.

Nitrogen fertiliser management in AFI, AFR and AEF led to an increase in total grain protein when compared to the absence of nitrogen fertiliser. According to Boehm et al. (2004), the increase in grain protein content is due to nitrogen, which is a basic constituent of proteins and amino acids (MALAVOLTA, 1981).

For the AFR management, the Fundacep 52 genotype showed a higher gluten strength than the other managements when ammonium nitrate was used, reaching 241 x 10^{-4} J, and could be classified in the bread class, according to Normative Instruction 38 (BRASIL, 2010). The behaviour of this character may be related to the greater availability of nitrogen in the grain formation process, increasing the quality and quantity of the protein, resulting in a strong gluten (FUERTES MENDIZÁBAL et al., 2010).

There were no significant differences in gluten strength between the nitrogen treatments for the Fundacep 52 genotype. According to Mendoza Duarte (2006), the application of urea under ideal conditions of air temperature, rainfall, air and soil humidity minimises the volatilisation of NH_3 and increases efficiency. The Tbio Mestre genotype showed higher magnitudes of gluten strength for the AFI and AEF managements when urea and ammonium nitrate were used. Guarienti et al. (2013) determined that this effect is due to the increase in gliadin and glutenin content in the grains, jointly increasing gluten strength. As for the sources, urea is superior for this trait in AEF management, as the nitrogen is made available more quickly and absorbed by the plant in the form of NO_3 or NH4 (CANTARELLA, 2007). Ammonium nitrate has this effect in the AFI and AFR managements, as it provides a slow release of nitrogen and reduces losses (MESQUITA, 2007).

The Tbio Itaipú genotype shows higher magnitudes of gluten strength for the AFR management using urea, and for the AEF management using ammonium nitrate, for this genotype the supply of nitrogen in the reproductive phase increases gluten strength. Gluten strength and protein content are parameters successfully used to select superior genotypes for industrial quality (KUCHEL et al., 2006). Ammonium nitrate was found to be superior to AFI and AEF. The Tbio Mestre genotype showed superiority for this trait.

Flour toughness shows that the Fundacep 52 genotype in the AUN and AFR managements has higher magnitudes for this character with the use of ammonium nitrate. The nitrogen sources revealed that urea was superior to AFI and AFR management, while ammonium nitrate was superior to AEF management. The Tbio Mestre genotype showed superior performance in terms of flour toughness for urea and ammonium nitrate in the AUN management. For this genotype, it can be seen that ammonium nitrate is superior in the AUN, AFI and AFR managements, and urea increases the character in the AEF management. For the Tbio Itaipu genotype, when urea was used, the AUN management was superior for toughness, and for ammonium nitrate, the AUN and AFI managements were superior. The Tbio Mestre genotype was superior to the other genotypes.

The Fundacep 52 genotype in the AFI and AEF managements proved to be superior to the others using urea. For ammonium nitrate, the superior management was AFR. Extensibility and protein content represent the ability of the dough to expand without breaking, the greater the extensibility the lower the flour yield (MÓDENES et al., 2009). For this genotype, urea proved to be superior for the AUN, AFI and AEF managements, while ammonium nitrate showed better results for the AFR management. The Tbio Mestre genotype showed superior extensibility to urea and ammonium nitrate in the AEF management. The Tbio Itaipu genotype was superior for extensibility in AEF management for both sources. Moraes et al. (2013) reported that increasing the dose of nitrogen improves alveography parameters.

The P/L ratio shows that the Fundacep 52 genotype was superior to the AUN and AFR managements using urea and ammonium nitrate. This response is linked to the content of glutenins and gliadins present in gluten, which increase the P/L ratio and predispose to an increase in high elasticity proteins with reduced extensibility, while a reduction in the P/L ratio indicates an increase in gliadins characterised as low elasticity proteins (MANDARINO, 1994). The Tbio Mestre and Tbio Itaipú genotypes have the highest P/L ratio due to the absence of nitrogen topdressing via urea and ammonium nitrate. Cazetta et al. (2008) show that increasing nitrogen increases the overall gluten strength and reduces the P/L ratio, thus improving the quality of flour for baking. Tbio Mestre and Tbio Itaipú were found to have a higher P/L ratio than Fundacep 52.

The Fundacep 52 genotype in the AFR and AEF managements had the highest number of drops (NQ) with urea, while for ammonium nitrate the superiority was emphasised in the AEF management. The supply of nitrogen promotes greater protein synthesis and minimises the effects of the a-amylase enzyme when there is excess moisture after the physiological ripeness of the grains (DERERA, 1989). According to Wieser et al. (2006), starch degradation, reserve proteins such as gliadins are rapidly degraded by a-amylase compared to glutenins. Ammonium nitrate was superior to urea in AFR management. The Tbio Mestre genotype in AFR management with urea and AFI management with ammonium nitrate showed a higher number of falls than the other treatments.

The Tbio Itaipú genotype was not significant with the addition of urea, but for ammonium nitrate the AFI, AFR and AEF managements showed a higher number of drops, the use of nitrogen in top dressing resulted in the highest number of drops, likewise Miranda et al. (2011), found that the availability of nitrogen in the soil did not imply a response to the technological quality of the wheat. The Fundacep 52 genotype shows that the use of urea provides superior wet gluten for AEF management. This is due to the fact that the supply of nitrogen in the grain filling phase results in greater synthesis of the gluten-forming proteins, gliadins and glutenins (FUERTES-MENDIZÁBAL et al., 2010). Ammonium nitrate for the AFI, AFR and AEF managements had a greater effect on this trait. Gluten is characterised by being a three-dimensional viscoelastic mass, which provides the characteristics of elasticity, plasticity and viscosity that are important for pasta (COSTA et al., 2008).

For the AFI, AFR and AEF managements, the Tbio Mestre genotype showed superiority for the wet gluten (GU) character when using urea, while ammonium nitrate was superior for AFI, AFR and AEF. With regard to sources, urea was superior in AFR management, as was ammonium nitrate in AEF management. The Fundacep 52 genotype was superior in terms of wet gluten compared to the other genotypes.

Technological quality parameters are influenced by management, nitrogen sources and genotypes. Nitrogen management with urea and ammonium nitrate did not influence protein content. The Tbio Mestre genotype has the highest gluten content, tenacity and extensibility when urea is applied at tillering and flowering, while the best responses to ammonium nitrate are obtained for the number of falls.

CHAPTER 6

An approach to improving dual-purpose wheat genotypes

Maicon Nardino, Ivan Ricardo Carvalho, Vinícius Jardel Szareski

Wheat *(Triticum aestivum* L.) is an important cereal in the agricultural, social and economic spheres, with average world production of around 650,000 tonnes for the 2012, 2013 and 2014 harvests, second only to maize (USDA, 2015). Countries such as Argentina, Australia, Canada and the United States are the main producers and exporters of wheat, while Brazil is the third largest importer of this cereal in the world. The soil and climate conditions allow wheat to be grown in a wide variety of environments and technological levels, and the southern region is the leading producer, with the state of Paraná being the largest Brazilian producer (CONAB, 2013).

The production chain for this cereal is essential, as grain production provides the basic raw material for making food products for humans and animals (MITTELMANN et al., 2000). In animal feed, wheat is characterised as a viable alternative to maize, as it is a highly available raw material with low added costs and high nutritional value. It can therefore be used to make feed and bran for cattle, pigs, sheep and poultry (MARQUES et al., 2007). These multiple uses are justified by the nutritional composition of its grains, which contain 87.77% dry matter, 54.93% starch, 11.49% crude protein and 2.37% crude fibre,

1.68% fat, 1.59% mineral material and 3819 Kcal Kg$^{'1}$ of energy (ROSAGNOetal., 2011).

On the other hand, advances in research into genetic improvement and crop management have made it possible to obtain dual-purpose wheat genotypes, where not only grain production is the goal, but also the ability to provide fodder for animals (BARTMEYER et al., 2011). By using dual-purpose genotypes, direct and indirect benefits are obtained, such as diversification of the farm, better use of physical space, reduction of periods with forage gaps and food shortages, crop-livestock integration and rational exploitation of the property (BORTOLINI et al., 2004).

In order to obtain genotypes that meet the dual-purpose requirements, breeders base themselves on the construction of a plant ideotype that shows rapid establishment, tillering potential, high dry mass production per unit area, tolerance to grazing and trampling, regrowth capacity, long vegetative period and short reproductive phase, high bromatological quality of the forage and adequate grain yield (MARTIN et al., 2010). In this way, dual-purpose wheat can be used for direct grazing of forage, production of silage, hay, pre-drying, and the grains can be used to produce energy for animals, mulch and green manure (FONTANELLI et al., 2009).

Currently in Brazil, the greatest use of dual-purpose wheat genotypes is concentrated in the southern region, but the large cultivated area reveals many producing micro-regions with specific soil and climate characteristics (FONTANELI, 2007). As a result, few genotypes are suitable for dual-purpose cultivation, and there is a growing need to meet producers' needs in terms of the availability of new genotypes. The complexity of the characteristics revealed in a dual-purpose genotype culminates in difficulties for the breeder, where a fraction is attributed to the need for the genotype to express superiority in terms of morphological, bromatological and grain yield components, and for these to supply the animals with energy (MARTIN et al., 2013).

Today, there are many doubts about the magnitude and meaning of the associations between these traits, as understanding these interrelationships helps to guide the appropriate selection strategy. According to Coimbra et al. (2000), understanding the association between traits is fundamental for breeders, as one trait can be responsible for the expression of another. On the other hand, the genetic improvement programme will be more successful by increasing the chances of gathering all the desirable characteristics in one genotype and ensuring that they meet the needs of producers in the most varied growing conditions.

The difficulties faced in selecting superior individuals force breeders to adopt differentiated and efficient alternatives. According to Carvalho et al. (2002), selection is made more difficult mainly due to actions resulting from the genotype x environment interaction (G x A), because when the character of interest is difficult to measure and it is not feasible to apply direct selection, it is determined which characters are associated with the expression of this character. In this way, indirect selection is used for secondary traits in order to achieve genetic gains for the main trait. Studies show that the oscillations in the genotype's response to different environments mean that indirect selection must be very well planned (CRUZ and REGAZZI, 1997). Indirect selection can be based on a character that is easy to measure, has high heritability and is associated with a character that has low heritability and is highly influenced by the environment (HARTWIG et al., 2007).

In order to better understand the associations between characters and direct selection, biometric models can be applied to genetic improvement and can be beneficial. In this way, canonical correlations make it possible to establish interrelationships between groups of characters involved in selection (TAVARES et al., 1999). According to Cruz and Regazzi et al. (1997) the use of this analysis allows breeders to understand the performance of more than one dependent character. Studies by Coimbra et al. (2000) reveal the great contribution of this method in selecting superior individuals. On the other hand, the *path analysis* developed by Wright (1921) allows total correlations to be broken down into direct and indirect effects, understanding the associations between the explanatory characters and the main character, and makes it possible to reveal the magnitude and direction of the associations. Studies by Kurek et al. (2001) show that this tool provides gains for plant breeding and is promising for determining the selection strategy, reducing the time taken to select genotypes, physical space and financial resources of the breeding programme. The aim of this literature review is to determine the biometric processes and models used to obtain and select dual-purpose wheat genotypes.

The origin of culture

Archaeological reports show that the first wheat crops date back 6,700 years BC in the Middle East, the ancient region of Mesopotamia comprising the vicinity of the Tigris and Euphrates rivers, today the territory of Iraq (TOMASINI and AMBROSI, 1998). Wheat is one of the pioneering crops in terms of domestication, and was the basis for feeding Asian, European and African civilisations. Its importance in the food sector has allowed this cereal to spread to various agricultural territories around the world (VESOHOSKI et al., 2011).

The genus *Triticum* has seven chromosomes in its basic genome. The best known species include diploid, tetraploid and hexaploid wheats, such as Triticum *monococum* L. (2n=2x=14), Triticum *turgidum L. (*2n=4x=28) and *Triticum aestivum L.* (2n=6x=42) respectively (GILL et al., 1991). In this way, common wheat *(Triticum aestivum* L.) is seen as a set of three complete diploid genomes, characterising an allopolyploid (AABBDD), where each genome comes from a species, namely *Triticum urartu (*AA), *Aegilops speltoides (*BB), and *Aegilops tauschii* (DD) (BRENCHLEY et al., 2012).

Because it is an allopolyploid, wheat shows polysomic inheritance for its characteristics, so a gene present in one of the genomes may be contained in the others (MORAES FERNANDES, 1982). Studies show that wheat's ability to adapt to different environmental conditions is due to the complexity of its genome (WALTER et al., 2009). This fact leads to peculiarities regarding the segregation pattern and the incorporation of genes, and genetic improvement aims to obtain agronomically suitable genotypes that will increase beneficial characteristics under biotic and abiotic stress conditions, better response to the genotype x environment interaction, and provide an increase in crop productivity (FEDERIZZI et al., 1999).

Botanical description

Common wheat is characterised as a grass belonging to the kingdom Plantae, division Magnoliophyta, class Liliopsida, order Poales, family Poaceae, genus *Triticum L.* and species *Triticum aestivum* L. (DEDECCA and PURCHIO, 1952).

The study by Slageren (1994) classifies six species of the *Triticum* genus: Triticum *urartu, Triticum monococcum, Triticum turgidum, Triticum timopheevii, Triticum zhukovskyi* and *Triticum aestivum.* There are subdivisions of the *Triticum aestivum* species: Triticum *aestivum* subsp. aestivum, *Triticum aestivum* subsp. *compactum, Triticum aestivum* subsp. *mancha, Triticum aestivum* subsp. *spelta* and *Triticum aestivum* subsp. *sphaerococcum.*

Morphological and physiological characteristics of wheat

The wheat crop has an annual cycle and plants with a cespitose growth habit, including genotypes with heights of 0.30 to 1.50 metres. The root system is characterised as fasciculate and can reach dimensions of 0.30 to 0.40 metres. The stem is classified as thatch and is made up of nodes and internodes, which are responsible for leaf insertion and thatch elongation, respectively. The leaf is composed of the sheath, which is characterised as an elongated structure attached to the thatch, the ligule is membranous and whitish, and the leaf blade is linear with parallel-inverted veins. The auricle is small to medium in size and may be hairy (FONTANELI et al., 2012).

Among the characteristics of wheat, tillering is extremely important to the productive scope of genotypes, as it provides compensatory effects through the emission of photosynthetically active offspring, which contribute assimilates to the main stem and form reproductive structures. According to Valério et al. (2009), the number of tillers per plant reflects the number of ears per unit area, the number and mass of grains per plant and consequently the grain yield of the genotype, as long as they are self-sufficient and do not act as a drain on the plant's assimilates. Wheat tillering can be controlled by genetic effects, hormonal effects through auxins and cytokinins, characteristics of the growing environment, water and nutritional demands, and the arrangement of plants in the canopy (VALÉRIO et al., 2009).

Thinning is essential for dual-purpose wheat because it increases the number of leaves, leaf area and forage production (MARTIN et al., 2010). According to Bortolini et al. (2004), it indirectly reduces lodging, apical dominance and plant height. Studies by Santos et al. (2011) show that the management of cuts in dual-purpose genotypes results in a stimulus in the emission of offspring, increases in leaf area, interception and utilisation of photosynthetically active radiation.

The periods of wheat development are divided into vegetative and reproductive, from seedling emergence to the appearance of the inflorescence, and then until physiological maturity, respectively (STRECK et al., 2003). The inflorescences are called ears and are formed by a set of spikelets attached individually to the node of the rachis. Studies show that wheat has a great capacity to modify the number of spikelets per spike through population management, nutrition and the genetic characteristics of the cultivar (TEIXEIRA FILHO et al., 2008).

There are three hermaphrodite flowers on the spikelet, two of which are viable and the central one is somewhat sterile. The male carpel of the flower shows three anthers and the female apparatus shows two stigmas and an ovary. The reproductive carpels are surrounded by the anthecium, which is formed by the palea and the lemma. After self-fertilisation, germination of the pollen grain, emission of the pollen tube and development of the ovary, the fruit called the caryopsis is formed, which is responsible for generating new individuals and propagating the species. Studies show that pollen viability and flower fertilisation are closely related to the growing environment, directly influencing the productivity and quality of the seeds produced (RIBEIRO et al., 2012).

Dual-purpose wheat

Dual-purpose wheat is characterised as a dual-purpose cereal because it provides fodder in the growing season and later provides grain for harvest (MARTIN et al., 2010). According to Hastenpflug (2009), the grains produced are destined for the production of bran and feed. On the other hand, studies by Ribeiro et al. (2010) show that dual-purpose wheat can reduce raw material shortages in the production of feed and efficiently replace maize, with wheat having a lower added price, high nutritional value and availability of grains between maize harvests.

Dual-purpose is a highly energetic alternative for the agricultural sector, as it is an energy source for animals in both the vegetative and reproductive periods. The fodder contains 23.00% crude protein, 53.00% neutral detergent fibre and 26.80% acid detergent fibre (FONTANELLI et al., 2009). In contrast, grains provide 87.77% dry matter, 54.93% starch, 11.49% crude protein, 2.37% crude fibre, 1.68% fat, 1.59% mineral material and 3,819 Kcal Kg^1 of energy (ROSAGNO et al., 2011).

This strategy makes it possible to successfully bring together activities aimed at different areas of agribusiness, so crop-livestock integration makes it possible to economically improve rural property. Studies by Del Duca et al. (2000) reveal the mutual benefits of crop-livestock integration for the soil, plants and animals, as well as increasing the producer's economic gain in the same physical space. Dual-purpose farming makes it possible to provide fodder directly to the animals, or to mechanise the production of pre-dried fodder, hay and silage, while the grain produced can be used as an energy source in animal feed (FONTANELI et al., 2009).

The southern region of Brazil has some peculiarities when it comes to forage production. Many of the species used have a fixed production period, which directly influences the scarcity of food for the animals. Studies show that the forage gap occurs in autumn and early winter (MEINERZ et al., 2011). In this way, dual-purpose wheat genotypes minimise the adverse effects of food shortages by enabling early sowing, rapid establishment, tolerance to trampling, high forage yields and the bromatological quality of the material produced (FONTANELI et al., 2009). Studies by Wendt et al. (2006) show that genotypes should have a long vegetative phase and allow for a greater number of cuts, while having a short reproductive period.

Wheat grown for dual purposes shows some care when it comes to plant management. When these are followed rationally, the activity results in success in both the productive and economic spheres. Well-used management techniques allow for an increase in the number of cuts, greater forage extraction and regrowth capacity. Studies show that the intensity of the stress caused to plants by the loss of leaf area can affect both forage and grain yields (BORTOLINI et al., 2004).

Therefore, some criteria should be considered, such as balanced fertilisation at sowing time and top dressing after each cut. Among the crop's needs is the addition of nitrogen, in order to boost regrowth and recovery of plants subjected to stress from defoliation. Studies by Sangoi et al. (2007) show that applying nitrogen to wheat increases the number of fertile shoots and grain yield. In dual-purpose cereals, nitrogen helps to establish leaf area and the crude protein content of the forage (DEL DUCA et al., 1999). Another important aspect is the time when the animals enter and leave for grazing, and the cutting height in mechanised management. If this is not taken care of, it can lead to damage to the plant's apical meristem, impairing elongation and grain yield (MARTIN et al., 2010).

Therefore, in order to achieve high fodder production in dual-purpose genotypes while minimising the effects of the fodder gap, sowing should be brought forward by 20 to 30 days, and the population density can be increased compared to traditional wheat (WENDT et al., 2006). The applicability of these management techniques depends on the growing environment, temperature, photoperiod, water supply, soil and production costs (MARTNetal., 2010).

The wheat genotypes recommended for dual purpose were developed from 2002 onwards through research by Embrapa Clima Temperado and Embrapa Trigo (WENDT et al., 2006). The main genotypes are BRS Figueira launched in 2002 from the Coker 762*2/Cnt8 cross, BRS Umbu launched in 2003 from the Century/BR 35 genotypes, BRS Guatambu and BRS Tarumã launched in 2004 from the Amigo/2*BR23 and Century/BR35 crosses, respectively, and BRS 277 launched in 2008 from the OR/Coker 93.33 genotypes (CAIRÃO et al., 2014).

Genetic improvement and obtaining genotypes

A wheat breeding programme aimed at obtaining new genotypes must be systematically organised and follow stages such as defining objectives, choosing genitors, hybridisation, forming segregating populations, choosing the method of conduction and selection, designing competitive trials, registering and protecting the genotype, multiplication and distribution of seeds. Obtaining a genotype designed for dual aptitude can be considered the primary objective of the plant breeder. In this way, the aim is to build a plant ideotype that combines high forage yield and bromatological quality, great tillering potential, tolerance to defoliation and trampling, and adequate grain yield, maximising the energy supply to animals (WENDT et al., 2006).

The selection of genitors is important for obtaining new genotypes, as it seeks genetic complementarity between genitors in order to increase the expression of agronomic characters of interest (PIMENTEL et al., 2013). Studies in wheat show that breeders can increase genetic variability by using cultivars in use, elite cultivars, *landraces* and breeding lines (SKOVMAND et al., 2006). After selecting the genotypes with the characteristics of interest and which meet the proposed objectives, the crossing blocks are organised, directing the crosses in such a way as to provide segregating populations with high expression of the target characters (SCHMIDT et al., 2009).

Wheat has hermaphrodite flowers, where cleistogamy provides self-fecundation through the flower's ability to fertilise itself before the petals open. Under these conditions, the breeder emasculates the flower and removes the three anthers, after which the inflorescence is protected. The anthers are then directed with viable pollen from the male parent and deposited on the stigma of the emasculated flower, identifying the cross with the appropriate information (ALLARD, 1971).

Viable crosses generate fully heterozygous individuals, which result in the formation of segregating populations. Studies show that the success of a wheat breeding programme is inextricably linked to the ability of the breeder to increase genetic variability in segregating populations resulting from crosses (PIMENTEL et al., 2013). According to Ramalho et al. (2012) the breeder's actions must be based on populations with high averages for the traits of interest, and also reveal high genetic variability.

Among the methods of selection and plant breeding in wheat genetic improvement with hybridisation, we highlight the use of mass selection, the population method, the genealogical method and *Single Seed Descent* (SSD). All of these methods are carried out from the second generation of offspring "F_2 " (BORÉM and MIRANDA, 2013).

Mass selection is less costly and requires less labour, and reveals a major contribution from natural selection through the conditions of the cultivation environment. The populations are sown at commercial densities from F_2 to F_4 , the best plants are harvested and mixed, and in the F_5 generation the sowing will be spaced out. Selection based on phenotype will give rise to the F_6 generation, in F_7 competition trials are set up (SCHEEREN et al., 2011).

The genealogical or *pedigree* method is based directly on the evaluation of the progenies, where the generated population is sown spaced apart and evaluated individually at F_2 , where superior individuals are selected and harvested separately. Each plant will form an F line$_3$, in F_4 selection is based not only on morphological characters, but on the characteristics of the progeny generated, and the selection of the best lines is repeated until homozygosity is achieved. Cultivation and use value (VCU) trials are then carried out with commercial witnesses, but the selected plants must be sown at the time and place where the future genotype will be used (BORÉM and MIRANDA, 2013). According to Allard (1971), this method is appropriate for qualitative characteristics determined by genes with large effects on the character.

The population method is based on growing F populations$_2$ under the same conditions as the future genotype, which must be harvested in *bulk*.

A sample of the previous generation is sown to make up the next generations from F_3 to F5. In F_6 the plants with promising phenotypes are sown in a line, and then the best Nails are harvested in *bulk* and submitted to the value for cultivation and use (VCU) trial with commercial witnesses (BORÉM and MIRANDA, 2013).

The *Single Seed Descent* (SSD) method is carried out by collecting just one seed from each individual F_2 . This seed will generate the individual of the next generations F_3 to F_5 and they will be grown spaced out. The best lines will be selected in F_6 and submitted to yield trials with other witnesses. This method allows generations to advance in the same year in the greenhouse, maintains the variability of the original population and increases homozygosity in each generation (BORÉM and MIRANDA, 2013).

Biometric models to identify characters of interest in selection

Canonical correlations aim to estimate the maximum correlation between groups of characters, since these characters tend to respond linearly (CRUZ et al., 2012). This analysis minimises the problems related to the presence of just one dependent character, as it does not distinguish which characters are dependent or independent, and reveals the maximum correlation between the groups (MORRISON, 1978).

Cruz et al. (2012) revealed that this method makes it possible to analyse the inter-relationships of groups with a varied number of characters, and the associations can be explained simply through a few correlations (CRUZ and REGAZZI, 2004). Associations between groups can be determined by the presence of at least two important characters (CRUZ et al,

2012). Where the number of canonical correlations is equal to the number of characters that form the smallest group, and the magnitude of these correlations *is* inversely proportional to the order in which they were estimated (CRUZ and REGAZZI, 2006).

Studies by Cruz et al. (2012) reveal that the statistical problem is linked to the maximum estimate of the linear correlation between groups, which promotes the determination of a weighting coefficient for each linear correlation of the characters. Estimating these coefficients for characters of interest to genetic improvement makes it easier to identify promising individuals by indirectly selecting characters that are easy to measure.

According to Carvalho et al. (2004), understanding the relationships between traits increases the efficiency of selection and reflects on the success of the genetic improvement programme. Studies by Santos and Vencovsky (1986) show that correlation makes it possible to identify associations between characters and reveal whether selection based on a particular character can have an effect on other characters. Canonical correlation is beneficial for genetic improvement and helps to understand the associations between groups of characters of agronomic interest (COIMBRA et al., 2000).

Path analysis was described by Wright (1921) and used in (1923) to provide a better understanding of the associations between characters, through the breakdown of simple correlations, and later detailed studies were carried out by Li (1975). This methodology makes it possible to quantify the magnitude and direction of associations between complex characters, revealing the importance of direct and indirect effects on the dependent character (CRUZ et al., 2012).

These effects are obtained through regression equations on standardised characters (CRUZ et al., 2006).

This analysis *is* characterised by conferring associations that make it possible to determine the interrelationships of cause and effect in the characters studied. In genetic improvement, it is commonly used to determine the importance of primary and secondary characters of the crop, and to guide the indirect selection of promising genotypes through characters of interest (CRUZ et al., 2012). According to Nogueira et al. (2012), understanding the associations between characters is essential for genetic improvement, as it helps to guide the selection process.

Path analysis has particular characteristics in that it is a regression coefficient and reveals a positive or negative direction. It is characterised by a standardised coefficient which allows characters measured in different physical units to be related, and does not express notations in its results (CRUZ et al., 2012). In genetic improvement, it is necessary to identify which characters show a high correlation with the main character, where the direct effect should be favourable to selection, while opposite directions between the total correlation and the direct effects indicate a lack of cause and effect association (CRUZ et al., 2006). Studies by Cruz et al. (2004) show that indirect selection is viable and can be used for characters that are difficult to measure, have low heritability and are highly influenced by the environment.

The definition of associations allows the breeder to understand the importance of each character in the expression of the others. Thus, path analysis aims to elucidate the relationships between characters, since indirect selection, when not based on the effects of the other characters, can reveal modifications in undesirable characteristics (SANTOS et al,

2000). Studies by Ramalho et al. (1993) show that correlation and associations between traits are important for plant breeding, as they determine the effect of selection on a particular trait, which will influence the other traits of the crop.

Final considerations

Dual-purpose wheat is an important alternative in the agricultural sector, as it enhances the energy dynamics of animal feed. Advances in genetics and plant breeding, as well as in crop management, have made it possible to develop genotypes that allow for both fodder and grain production. This strategy allows producers to increase their farm income, integrate crop-livestock activities, reduce the effects of the forage gap, maximise the physical space of the property, benefit the environment-plant-animal dynamic and economically increase farming activity. The benefits of using dual-purpose genotypes are justified by their wide-ranging uses, where they replace winter grasses and have higher energy yields, both for fodder and grain, which is rich in starch and is used to make bran and feed.

Research has developed genotypes recommended for broad regions, so there is a lack of specific genotypes for micro-regions and different technological levels of management, meeting the quantitative and qualitative demands for fodder and grains, increasing the energy supply in farming. Therefore, a genetic improvement programme must address effective strategies to identify promising individuals and enable genetic gains for the crop. Today, genetic improvement with an emphasis on dual-purpose genotypes does not give the necessary importance to secondary traits, and there is a lack of detailed identification of the interrelationships between forage, bromatological and grain yield traits. Knowing these associations through biometric models, identifying the magnitude and direction of the relationships between characters, makes it possible to guide the selection strategy and minimise the expenditure of time and financial resources.

CHAPTER 7

Strategic positioning of wheat genotypes for seed production

Vinícius Jardel Szareski, Ivan Ricardo Carvalho, Maicon Nardino

Wheat *(Triticum aestivum* L.) is a cereal belonging to the Poaceae family, used as a raw material in the production of bread, pasta and biscuits. Worldwide wheat production is close to 700 million tonnes, with Brazil producing six million tonnes a year. Of the total national production, the states of Rio Grande do Sul and Paraná contribute the most to Brazilian production (CONAB, 2015). In recent years, triticulture has expanded to the Cerrado regions, which has contributed to an increase in Brazilian production (MELEIRO et al., 2013).

Grain yield in wheat *is* the result of the joint action of several traits of agronomic interest, which are determined by genetic and environmental factors and the genotype x environment interaction. Thus, the effects of the growing environment are the most pronounced for the wheat crop, directly influencing the productive performance of the genotypes, the period from flowering to plant height and the specific mass of the grains (CARGNIN et al., 2006). Thus, the genotype x environment interaction stems from the differential responses of each genotype to changes in the biotic and abiotic characteristics of the environment (YAN & HOLLAND, 2010).

The strategies used to grow wheat must be supported by the choice of genotype, sowing time and cultivation,

The soil and climate characteristics of the environment, temperature and solar radiation influence crop productivity (SUBEDI et al., 2007). Research shows that planning and management techniques, when used properly, result in an 80 per cent increase in wheat productivity, especially when the crop *is* sown at the right time (BASSOI et al., 2005; COVENTRY et al., 2011).

Agronomically suitable genotypes when grown in favourable environments, with less pressure from insect pests and diseases, can guarantee producers uniform plant *stands* after emergence, increased productivity, vigorous seeds and high physiological quality (CARVALHO & NAKAGAWA, 2000). Thus, qualitatively superior seeds tend to generate more vigorous seedlings with an appropriate *stand* and rapid establishment in the field. The physiological potential of seeds is attributed to a set of genetic, physical, physiological and health characteristics (HUBBARD et al., 2012).

Among the factors that influence obtaining high quality seeds are the sowing time, harvesting process, weather conditions, processing and seed storage (MARCOS FILHO, 1999). In this sense, it is necessary to understand the agronomic performance of genotypes and determine the genetic variation between available genotypes using the generalised Mahalanobis distance (CRUZ et al., 2012).

The aim of this study was therefore to assess the effect of sowing time on agronomic and physiological components, in order to identify the period that enhances the increase in components; and to analyse the dissimilarity and relative contribution of characters in elite wheat genotypes using multivariate techniques.

The experiment was conducted during the 2014 agricultural season in Tenente Portela - RS, at the geographical coordinates of 27°22'28"S and 53°45'28"O, with an altitude of 501 metres. The climate, according to Kõppen, is characterised as subtropical *Cfa,* and the soil is classified as typical Latossolo vermelho alumino férrico (STRECK, 2008).

The experimental design used was randomised blocks, organised in a factorial scheme, with 11 wheat genotypes x 2 sowing times, arranged in three replications. The genotypes used were: BRS Parrudo, TBIO Sintonia, TBIO Mestre, TBIO Iguaçu, TBIO Itaipu, Mirante, TBIO Sinuelo, Quartzo, BRS 331, BRS 327 and TBIO Pioneiro. The sowing times were: Season I: sown on 01/05/2014, and Season II: sown on 01/06/2014.

The experimental units consisted of ten rows five metres long and spaced 0.17 m apart. Sowing was carried out using a tractor-seeder combination in a direct sowing system, with a population density of 330 viable seeds per square metre for all genotypes. The base fertiliser was 350 kg ha^1 of N-P-K in the 10-20-20 formulation, and 150 kg ha[-1] of nitrogen in the amide form was used as top dressing, applied at full tillering. At both sowing times, weeds, insect pests and diseases were controlled preventively in order to minimise the biotic effects on the experiment.

The traits of agronomic interest were measured on plants in the six central rows of the experimental unit, where one metre from each end was disregarded. Ten representative plants were then selected from each experimental unit. The yield characteristics were determined using the methodology proposed by Carvalho et al, (2015) and these were plant height (AP, in centimetres); ear mass (ME, in grams); ear length (CE, in centimetres); number of grains per ear (NGE, in units); ear insertion height (AIE, in centimetres); grain mass per ear (MGE, in grams); hectolitre weight (PH, in kg hl[-1]); thousand grain mass (MMG, in grams); grain yield (RG, in kg ha[-1]), with humidity corrected to 13%. The physiological characteristics followed the methodology proposed by the Seed Analysis Rule (BRASIL, 2009), namely: first germination count (PCG, in %); germinated seeds (SG, in %); radicle length (CR, in centimetres); aerial part length (CPA, in centimetres); seedling dry mass (MSP, in grams).

The analysis of variance revealed a significant interaction ($p<0.05$) between wheat genotypes and sowing times for the following traits: ear length (EC), ear mass (EM), number of grains per ear (NGE), grain mass per ear (GEM), thousand grain mass (GGM), grain yield (GY), first germination count (FGC), germinated seeds (SG), seedling dry mass (SDM), aerial part length (APL) and radicle length (RL). No significant interaction was observed for the characters plant height (PH) and ear insertion height (EIA). Hectolitre weight (HW) showed no significant differences for the variation factors. The coefficients of variation were low, ranging from 0.68 to 16.45%, thus characterising adequate experimental precision with reliable results.

Plant height (AP) and ear insertion height (AIE) were not influenced by the sowing times, but differences were obtained for the genotypes, where TBIO Itaipu, Mirante, TBIO Sinuelo, Quartzo, BRS 331 and TBIO Pioneiro were superior to the other genotypes for the characters in question. The ear length (EC) for sowing time (I) showed that the BRS 331, Quartzo, Mirante and TBIO Sintonia genotypes were superior to the others. Sowing time (II) showed that the genotypes BRS Parrudo, TBIO Sinuelo, Quartzo and BRS 331 were superior. When relating the sowing times, the BRS Parrudo genotype showed significant and superior differences for time II and demonstrated better behaviour with later sowing times, while the Mirante genotype showed the opposite behaviour. Larger ears tend to increase the number of spikelets and consequently provide a higher number of grains and yield per plant. Research has shown that higher ECs alter the translocation of assimilates in the ear and result in oscillations in the quality and mass of the grains in the ear (ANDERSSON et al., 2004; FIOREZE & RODRIGUES, 2014).

For the ear mass (EM) character, the genotypes TBIO Sintonia, TBIO Iguaçu, Mirante, TBIO Sinuelo, Quartzo, BRS 331 and TBIO Pioneiro showed superiority for sowing time (I), while for sowing time (II) the genotypes TBIO Sintonia, Mirante and TBIO Pioneiro showed the lowest magnitudes for this character. Among the sowing times, the genotypes TBIO Sintonia, TBIO Iguaçu, Mirante, TBIO Sinuelo, Quartzo and TBIO Pioneiro showed the highest ME when sown at time I, making them suitable for earlier sowings. The differences observed between the variation factors indicate that wheat ear mass is influenced not only by the intrinsic characteristics of the genotype, but also by the environmental conditions portrayed by the sowing times. In this way, environmental conditions are decisive for the performance of genotypes (YAN & HOLLAND, 2010). Studies by Silva et al. (2011) show that the sowing time influences the behaviour of genotypes due to the environmental changes that occur during this period.

The number of grains per ear (NGE) was superior for the le II sowing times for the Quartzo and BRS 331 genotypes. The results obtained for this trait indicate that these genotypes have the highest magnitudes for NGE regardless of the sowing time tested. Differences in the performance of this trait were obtained for the TBIO Sintonia genotypes, where sowing season I increased NGE, while the BRS *3T1* genotype showed better results when sown in season II.

Grain mass in the ear (GGM) is an important yield component for wheat crops, where sowing season I indicated that the TBIO Sinuelo genotype was superior to the others; under these conditions, the BRS 327 genotype performed less well. Later sowing in season II resulted in better performance for TBIO Sinuelo, while TBIO Sintonia had the lowest magnitudes of this trait. In relation to sowing times, it was observed that the genotypes TBIO Sintonia, TBIO Iguaçu, Mirante, TBIO Sinuelo, Quartzo and TBIO Pioneiro showed the best results for MGE when sowing was done earlier in season I, however, only the genotype BRS 327 showed better results when sowing was done later (season II).

Grain yield in wheat is determined by the number of ears per unit area, number of grains per ear and grain mass, so highly productive genotypes tend to express high magnitudes of these traits (BRANCOURT-HULMEL et al., 2003). On the other hand, increasing the number of grains in the ear modifies the dynamics of assimilate partitioning and the direction of these to the grains, thus resulting in lighter grains with lower protein proportions (ASHRAF TAJAMMAL et al., 2003).

The grain yield (GY) for sowing season I showed superiority for the genotypes TBIO Sintonia, TBIO Iguaçu, Mirante, TBIO Sinuelo, BRS 331 and Quartzo, while the lowest GY was observed for the genotype BRS 327. Sowing time II shows the least variation in this trait among the genotypes, with lower magnitudes for the TBIO Sintonia and TBIO Pioneiro genotypes. When comparing RG performance between sowing seasons, the genotypes TBIO Sintonia, TBIO Iguaçu, TBIO Sinuelo, Quartzo and TBIO Pioneiro are more productive when sown in season I, while the BRS 327 genotype shows better results in season n.

The answers obtained in this study allow us to infer that higher yields can be obtained when sowing is earlier, allowing genotypes to have their initial establishment period at lower temperatures, enhancing tillering, a short vegetative period and a long reproductive period, where the increase in fertile tillers and ears per plant leads to an increase in the number and mass of grains, which is directly reflected in the grain yield of wheat. Research carried out by Coventry et al. (1993) and Sial et al. (2005) corroborates the results obtained, where sowing wheat at later times reduces productivity.

The mass of a thousand grains (MMG) can be considered a character indicative of the most productive genotypes, so the genotypes BRS Parrudo, TBIO Sintonia, TBIO Mestre, TBIO Iguaçu and TBIO Sinuelo were superior for sowing time I, while the genotypes TBIO Itaipu, Mirante, Quartzo, BRS 331 and BRS 327 showed less evidence of this character. The performance of the MMG trait for sowing time II was higher for the TBIO Sinuelo genotype, which is characterised by having a medium to late cycle, with a longer period for the synthesis, accumulation and remobilisation of photoassimilates for the grains, which consequently results in larger grains and greater mass (SUPRAYOGI et al., 2011).

Among the sowing times, it was observed that the genotypes BRS Parrudo, TBIO Sintonia, TBIO Mestre, TBIO Iguaçu, TBIO Itaipu and Quartzo showed the highest MMG for sowing time I. Research has shown that grain mass in wheat is directly dependent on the characteristics of the genotype, nutritional management, water supply and the conditions of the growing environment (SIMMONDS et al., 2014).

The first germination count (PCG) provides a preliminary indication of the vigour attributed to the seeds, where sowing time (I) determines that only the TBIO Pioneiro genotype differs from the other genotypes tested, and reveals that the seeds produced under these conditions are less vigorous. Sowing time (II) reveals superiority for the genotypes TBIO Mestre and BRS 327, while inferiority for this character is observed for the genotypes BRS Parrudo, TBIO Sintonia and TBIO Sinuelo.

In relation to sowing times, the genotypes BRS Parrudo, TBIO Sintonia, Mirante and TBIO Sinuelo showed different behaviour to the other genotypes, with more vigorous seeds when sown in season I. Bringing forward the sowing period for wheat allows the plants to establish, grow and develop properly. By intercepting, absorbing and assimilating photosynthetically active solar radiation, it allows for greater synthesis and accumulation of assimilates in the plant tissue, so that during the reproductive period these will be directed to the seeds and will lead to an increase in physiological quality. The physiological potential of seeds depends on field cultivation techniques, harvesting, processing and storage processes (COSTA et al., 2005).

The germinated seeds (SG) trait *is a* standard and reliable indicator of the physiological quality of the seeds, where sowing time I indicates that the genotypes TBIO Sintonia, TBIO Mestre, Mirante and BRS 327 were superior, however, low magnitudes of this trait are evident in the genotypes TBIO Iguaçu and TBIO Pioneiro. Sowing time (II) revealed that the genotypes TBIO Mestre and BRS 327 showed the highest magnitudes for the SG trait, while low performance was expressed by TBIO Sintonia, TBIO Sinuelo and TBIO Pioneiro. The physiological characteristics of the seeds were not satisfactory for some genotypes at sowing time II, where the percentage of germinated seeds was lower than that required by current legislation (Law n°10.711), which determines SG=80% as a minimum (BRASIL, 2009).

With regard to sowing times, the genotypes TBIO Sintonia, Mirante, TBIO Sinuelo and TBIO Pioneiro showed the best results for the physiological quality of the seeds produced when sown in season I. Research by Viganó et al. (2010) has shown that different crop seasons and sowing times in wheat directly influence the physiological quality of the seeds produced, regardless of the genotype; therefore, earlier sowings resulted in qualitatively superior and vigorous seeds. Medina et al. (1997) & Sá et al. (1997) state that the physiological quality of wheat is extremely dependent on the characteristics of the growing environment.

Seedling dry mass (SDM) is used to express the vigour attributed to the seeds, where sowing time (I) shows that the genotypes BRS Parrudo, TBIO Sintonia, TBIO Mestre, TBIO Itaipu, Mirante, TBIO Sinuelo, Quartzo, BRS 331 and BRS 327 showed similar but superior responses and have more vigorous seeds than the other genotypes. However, sowing time (II) reveals that only the genotypes TBIO Mestre, BRS 327 and TBIO Pioneiro are the most vigorous.

The magnitudes obtained for MSP indicate that the vigour attributed to wheat seeds is more related to the sowing season or growing environment than to the characteristics of the genotype. This is justified by the superiority of season I for the BRS Parrudo, TBIO Sintonia, TBIO Iguaçu, Mirante, TBIO Sinuelo and Quartzo genotypes. The differences between sowing times may be due to unfavourable weather conditions, high rainfall, high relative humidity and temperature during flowering, physiological ripeness and harvest (VIGANÓ et al., 2010).

The length of the aerial part (CPA) makes it possible to understand the initial growth of the seedlings, where sowing time I indicates that the genotypes TBIO Iguaçu, TBIO Sinuelo, BRS 327 and TBIO Pioneiro were superior. These results can be related to genotypes that, as well as being vigorous, have a quick initial start and establishment of the seedlings in the field, and can be more competitive interspecifically. Sowing time (II) shows that the genotypes BRS Parrudo, TBIO Sintonia, TBIO Mestre and TBIO Pioneiro stand out when grown late in the season. Only the TBIO Iguaçu genotype shows differences in CPA due to sowing time.

The length of the radicle (CR) was superior for the TBIO Mestre genotype at sowing time (I), but for time (II) the best results were obtained for the TBIO Itaipu and BRS 327 genotypes.

The CR showed a high influence of growing conditions, as the genotypes BRS Parrudo, TBIO Mestre, Mirante, TBIO Sinuelo and BRS 331 showed the best results when sown in season (I), while only TBIO Itaipu showed a higher CR for season (II).

This trait can also be considered as an indicative parameter of seed vigour, where the seeds tend to be more vigorous and have more reserves in the endosperm, thereby enhancing the emission of root primordia. Thus, superior genotypes for this trait have a greater capacity to form a root system that is better able to explore the soil, absorb water and nutrients, as well as helping to support the plant.

In order to determine the genetic dissimilarity between the wheat genotypes tested, the Generalised Mahalanobis Distance was used using the UPGMA grouping method, where the 14 characters measured were used to make up the distance matrix. The average genetic distance (21.705) obtained by measuring the characters was used as the criterion for separating the groups in the dendrogram, and is represented by the dashed line, which resulted in the formation of two large groups of genotypes.

Group (G I) is made up of just two genotypes, the cultivars TBIO Mestre and BRS 327, which are more distant from the others. Group (G II) is made up of the remaining genotypes, but the general average of the distances allowed it to be subdivided into a small subgroup made up of the genotypes TBIO Sinuelo and TBIO Pioneiro, and another subgroup made up of seven genotypes: TBIO Sintonia, TBIO

Iguaçu, BRS Parrudo, TBIO Itaipu, Mirante, Quartzo and BRS 331. The different results obtained in this study in terms of genotype responses can be explained by the dissimilarity between the genotypes, which can be seen in the formation of three groups.

The relative contribution of the characters was obtained using Singh's method (1981) and reveals which characters are decisive in distinguishing the genotypes tested, out of the 14 characters measured in the 11 wheat genotypes. It was observed that the distinction between the genotypes is supported by four characters which together contribute 61.8 per cent. In isolation, it is clear that ear insertion height (AIE) is responsible for 28.07 per cent, germinated seeds (SG) for 12.36 per cent, number of grains per ear (NGE) for 10.59 per cent and plant height (AP) for 10.15 per cent.

The use of the relative contribution made it possible to define that ear insertion height (AIE), germinated seeds (SG), number of grains per ear (NGE) and plant height (AP) are essential characters to be measured in the wheat crop in order to be able to discriminate between the tested wheat genotypes.

Sowing wheat in May (season I) favours an increase in ear mass, ear grain mass, grain yield, thousand grain mass, first germination count, germinated seeds and seedling dry mass in the genotypes tested compared to sowing in June (season II).

Dissimilarity using the generalised Mahalanobis distance allows the genotypes to be divided into two large groups, with TBIO Mestre and BRS *3T1* being the most dissimilar genotypes.

The characters with the greatest relative contribution to discriminating between wheat genotypes are: ear insertion height, germinated seeds, number of grains per ear and plant height, which explain 61.8 per cent of the total variation.

CHAPTER 8

Morphological characteristics and seed production in dual-purpose wheat

Ivan Ricardo Carvalho, Vinícius Jardel Szareski, Maicon Nardino

Dual-purpose wheat *(Triticum aestivum* L.) is a cereal characterised by its ability to provide fodder during the growing season, while also enabling grain production (MARTIN et al., 2010). It is an excellent alternative for dealing with periods of fodder scarcity and for economically optimising farms in southern Brazil (MEINERZ et al., 2011). Its importance is due to its use in traditional pastures, pre-drying, hay and silage (FONTANELI et al., 2009).

The diversity of commercially available dual-purpose wheat genotypes suitable for southern Brazil is limited. Among the genotypes, BRS Figueira, BRS Umbu, BRS Guatambu, BRS Tarumã and BRS 227 can be highlighted (CAIRÃO et al., 2014).Dual-purpose wheat genotypes must have the following characteristics: rapid field establishment, intense tillering, high dry matter production, tolerance to grazing or cutting, good bromatological quality and high grain yield (MARTIN et al., 2010; CARVALHO et al., 2015).

Due to the variability of the country's soil and climate conditions, the genotype-environment interaction provides different responses from genotypes when subjected to different environmental conditions, altering their performance and causing a reduction in the relationship between genotype and phenotype (YAN & HOLLAND, 2010). In this context, the aim is to increase genetic variability for characters of agronomic interest that can make genotypes more adapted and stable to adverse conditions.

The chemical agent *Ethyl Methane Sulphonate,* when applied to seeds, can help to obtain gene, morphological and physiological changes (BORÉM & MIRANDA,

2009; MOTURI & CHARYA, 2010). This technique is used in wheat cultivation to assess the damage produced by different concentrations on the physiology of hexaploid plants (SILVA, 1998). In oats, to analyse changes in genetic variability in the plant height character, as well as to create genetic variability for the vegetative cycle character (COIMBRA, 2004, 2005), as well as to compare the efficiency of artificial crosses *versus* chemical mutagens in the plant height character in oats (OLIVEIRA, 2012). In barley, it can be used to analyse the cytological effects of pyrethroid insecticides (KARNOPP, 1999).

Recently, more than 2,500 genotypes have been released using this technique (AHLOOWALIA et al., 2004).Given the lack of research involving chemical agents in the cultivation of dual-purpose wheat.

The aim of this study was to evaluate the application of different doses of *Ethyl Methane Sulphonate to* double-purpose wheat seeds and their response to morphological characters when measured at different evaluation periods.

The experiment was carried out during the 2013 agricultural harvest in the experimental field of the Federal University of Santa Maria, Frederico Westphalen *Campus,* RS. The coordinates correspond to latitude 27°23'26"S and longitude 53°25'43"O, with an altitude of 490 metres. The soil is classified as Latossolo vermelho alumino fémeo (EMBRAPA, 2006), and the climate is characterised by the Kõppen classification as *Cfa* of the subtropical type (BERNARDI et al., 2008).

The experimental design was randomised blocks, with two dual-purpose wheat genotypes x five doses of *Ethyl Methane Sulphonate and* four evaluation periods, and four replications.

Sowing was done using a direct sowing system, with a population density of three million seeds per hectare. The base fertiliser was 250 kg ha'^{l} of NPK in the 05-20-20 formulation, and 100 kg ha' of N was applied as top dressing at the full tillering stage. The nitrogen source was urea at 45% nitrogen and the experimental units were made up of 12 rows 2.0 metres long with a spacing of 0.17 metres.

The genotypes used were: BRS Tarumã and BRS Umbu. The doses of *Ethyl Methane Sulphonate* used corresponded to OmL Kg' ^O^SmL Kg'^{l} ; 0.50mL Kg'^{l} ;

0.75mL Kg^{-1} ; 1.0mL Kg^{-1} of seeds and the evaluation periods corresponded to the phenological stages: \(*Feeks 3 - tillering*); \\(*Feeks 8 - rubberising*); III(FeeFs' *10.5.1 - flowering*) and IV(Feeks *11.1 - grain filling*).

The seeds were placed under a nylon mesh where they remained for two hours with the *Ethyl Methane Sulphonate* treatment at the established doses. After being removed from the treatments, the seeds were washed under running water for one hour.

The characters were measured by sampling ten random plants in each experimental unit, and the leaf area was determined using a LI-3000® leaf area integrator, with the results expressed in square centimetres (cm^2). Root diameter was measured using a digital caliper, and the average root diameter was measured, the results in millimetres (mm).The number of fertile tillers was determined as a direct percentage of the number of tillers with viable ears in one linear metre, the results in units.Plant height, measured from ground level to the apex of the main ear and disregarding the aristas, the results expressed in centimetres (cm). Diameter of the culm of the offspring: the average diameter of the culm of the offspring was measured using a digital caliper, with the results expressed in millimetres (mm).Root length was determined by measuring from the neck of the plant to the end of the roots, results in millimetres (mm), while the chlorophyll index was measured using a SPAD-502® chlorophyll meter, based on measurements on 50 leaves per experimental unit, always at 2pm, on the first leaf after the flag leaf, fixing the chlorophyll meter in the middle third of the leaf blade (results in mg m').^{2}

The analysis of variance revealed a significant interaction between dual-purpose wheat genotypes x doses of *Ethyl Methane Sulphonate* x evaluation periods for leaf area and root diameter. There was an interaction between dual-purpose wheat genotypes x evaluation periods for the characters number of fertile tillers, plant height, tiller diameter and chlorophyll index. There was no interaction for the root length character.

The leaf area trait is directly related to maximising the interception of incident solar radiation, which can be reflected in an increase in the crop's growth rate, an increase in the mass of a thousand seeds and grain production (OKUYAMA et al.,

2004). It was observed in period I (tillering) that the different doses of *ethyl methane sulfonate had* little influence on the increase in the character for both cultivars. In evaluation period II (germination), the BRS Tarumã cultivar showed the greatest increase at a dose of 1.0 mL kg^1 , while the BRS Umbu cultivar only showed a positive response up to a dose of 0.75 mL kg^1 . In period III (flowering) there was no increase in the character as a function of the doses of *Ethyl Methane Sulphonate.* For period IV (grain filling), the dose of 0.75 ml kg$^{'1}$ showed an increase for both cultivars (Figure 1-A).

When comparing evaluation periods (Table 1), it can be seen that the greatest magnitude was revealed in period III (flowering). Thus, leaf area is of great importance as an active photosynthesising tissue, providing greater partitioning of assimilates during grain filling (SILVA et al., 2003). Therefore, the greater availability of assimilates close to anthesis could mean more fertile flowers and, consequently, a greater number and size of grains (RODRIGUES, 2000; SILVA et al., 2003).

For the root diameter character, as a function of the doses of the agent *Ethyl Methane Sulphonate,* it was observed that period I (tillering) did not lead to an increase for the cultivars. In period II (budding) there was a slight reduction in magnitude for both cultivars. In period III (flowering), the BRS Tarumã cultivar showed a reduced increase with the dose of 1.0 mL Kg$^{'1}$, while the BRS Umbu cultivar showed a reduction. In period IV (grain filling) there was a slight increase for both cultivars.

When comparing the evaluation periods for the root diameter character (Table 2), there was an increase for the BRS Tarumã cultivar in periods II (rubberisation) and III (flowering), but the BRS Umbu cultivar showed the greatest magnitude in periods II (rubberisation) and IV (grain filling).

It was observed that the number of fertile offspring, as a function of the doses of the agent *Ethyl Methane Sulphonate,* did not show a significant increase. The highest magnitudes were seen in period III (flowering) for both cultivars, with the greatest intensity of the character revealed in the BRS Tarumã cultivar.

The number of fertile offspring character contributes to the number of ears per unit area and consequently increases grain yield. This trait depends on the intrinsic characteristics of the genotype, the management applied, the growing environment and the genotype x environment interaction (VALÉRIO, 2008, 2013; OZTURK et al., 2006).

Plant height is highly important, as tall cultivars are more susceptible to lodging and consequently to significant losses in production potential. Shorter plants have a greater capacity to intercept solar radiation during critical periods of production definition, as well as directing the carbon not used in height growth (CHAVARRIA et al., 2015). Increasing the doses of the agent *Ethyl Methane Sulphonate* did not interfere with the character in question. Evaluation period IV (grain filling) showed an increase in both cultivars.

There was no increase in the diameter of the fertile daughter stalks as a result of increasing doses of *Ethyl Methane Sulphonate.* Evaluation period IV (grain filling) was superior for both cultivars, with the highest magnitudes being observed in the BRS Umbu cultivar.

Studies by Martin et al. (2010) showed that the yield components of dual-purpose wheat are largely influenced by tillering, water availability, nutrition, quantity and quality of light, temperature and number of cuts. According to Simmons et al. (1982), cultivars that express greater earliness potential show a reduction in the diameter of the stalks.

For the chlorophyll index in relation to the doses of *Ethyl Methane Sulphonate,* there was no increase for the character as the doses increased. Chlorophyll content was superior in period III (flowering) for the BRS Tarumã cultivar and in period II (budding) for the BRS Umbu, with little variation in magnitude between cultivars.

In wheat growing, the chlorophyll index is widely used to predict the need for nitrogen fertilisation, as the amount of this pigment correlates positively with the nitrogen content in the plant (SENA JÚNIOR et al. 2008). The similarity between the cultivars for this trait may be related to the fact that they were treated similarly.

In relation to root length as a function of *Ethyl Methane Sulphonate* doses, there was a reduction in root length as the dose increased, revealing the phytotoxic effect of the mutagen on the root growth of dual-purpose wheat. A larger size or volume of roots allows the soil to be better occupied, favouring the absorption of water and minerals for plant growth and development (BARBER, 1988).

Evaluation period II (rubberisation) gave the greatest root length, which did not differ from evaluation period III (flowering). As for the genotypes, BRS Umbu showed superiority in terms of root length compared to BRS Tarumã.

This character is influenced by soil pH, exchangeable aluminium content, overall density, water storage and hydraulic conductivity. It can also be compromised by management, soil compaction, chemical element toxicity and excess water (MÚLLER et al., 2001; VALADÃO et al., 2015).

The dual-purpose wheat cultivars behaved differently from one another and between the evaluation periods when the seeds were subjected to the application of the mutagen *Ethyl Methane Sulphonate*.

Increasing the doses of the agent *Ethyl Methane Sulphonate* reveals variability for the characters leaf area, root diameter, number of fertile tillers, tiller diameter and chlorophyll index; however, it has a negative effect on the length of the main root.

CHAPTER 9

Alternative fertilisation for wheat seed production

Maicon Nardino, Ivan Ricardo Carvalho, Vinícius Jardel Szareski

Wheat *(Triticum aestivum* L.) comes from south-east Asia, belongs to the *Poaceae* family and is one of the most important cereals in the world economy, with annual production of 600 million tonnes. The world's main producers and exporters are the European Union, the United States and China, and importers are Brazil, China and India. In Brazil, the southern region is the leading producer of the cereal in the winter period, producing 45.5 per cent of total Brazilian production (CONAB, 2016).

In recent years, new sustainable strategies have been sought in the agricultural sector, especially in relation to the destination of surplus waste from animal production systems, such as poultry litter, which *is* produced in large quantities in poultry houses. The appropriate use of this waste for fertiliser can bring about a significant increase in productivity, as well as significantly reducing the negative effects on the environment. The lack of symbiosis with microorganisms means that grasses such as wheat respond positively to an increase in nitrogen in the soil solution. Fertilising with poultry litter as a source of nutrients can replace all or part of the use of chemical fertilisers (MENEZES, J.F.S, 2004).

Wheat grains have excellent bromatological quality, with a high concentration of proteins. It is considered a medium-investment crop, with a high risk during its development period, caused mainly by frost, hail and excessive rainfall (Wendt, 2007).Nitrogen fertilisation is necessary for this crop. According to Frank & Bauer (1996), the availability of nitrogen during the period of cell differentiation is highly related to grain yield, as this period is when the ear develops and spikelets are formed. The same has been seen for other winter cereals, where the application of nitrogen in top dressing at the tillering stage provides an increase in grain yield

(PERUZZO, 2000). However, the plant's high need for this nutrient results in high production costs. According to Ragagnim (2013), most of the fertilisers used in Brazil are imported, which increases costs.

The growing consumption of poultry meat is important for the economy, especially in the states of Santa Catarina, Paraná and Rio Grande do Sul. Poultry farming generates a large amount of waste, which can be used as fertiliser due to its concentration of nutrients and can be used as an alternative organic fertiliser for agricultural crops. According to Avila et al. (2007), poultry litter can be used as a source of nutrients for fertilising crops. Silva et al. (2011), when using poultry litter as a nitrogen source, observed an increase in the morphological parameters of maize plants.

Poultry litter can be characterised as waste made up of sawdust, wood shavings, rice husks, poultry droppings, feathers and feed waste, among others. This organic waste is rich in nutrients, including nitrogen, most of which is in organic form and requires a period of mineralisation to be made available for absorption (SANGOY et al., 2008). The aim of this study is therefore to assess the possibility of using poultry litter as an alternative fertiliser for wheat crops.

Wheat belongs to the *Poaceae* family and the *Triticum* genus, which contains around 30 species. This genus has seven chromosomes in its genome. The best known species are diploid, tetraploid and hexaploid wheats: Triticum *monococcum* has 14 chromosomes, Triticum *durum* has 28 chromosomes and *Triticum aestivum, known as* common wheat, has 42 chromosomes and is considered to be the species of greatest commercial interest (BRAMMER, 2001). The complexity of its genome is what gives wheat cultivars wide adaptation, making it possible to grow this cereal in a variety of environments (WALTER et al., 2009).

This crop was among the first to be domesticated by man between 7000 and 9000 BC in Southeast Asia, and was later introduced to India, China and Europe. In Brazil, it was introduced by the colonisers in the 11th century and was the first agricultural activity in Brazil. The first Brazilian crops were grown in the state of São Paulo, and later in Rio Grande do Sul, which today, together with the state of

Paraná, accounts for the majority of Brazilian production.

Advances in research have brought new genotypes to the market that are better adapted to the regions, making its cultivation a lower risk activity, with cultivars that are better adapted and more stable. It has thus become the main winter cereal in the southern region and is exported to different countries.

Worldwide, wheat ranks first in terms of production volume. In Brazil, annual production fluctuates between 5 and 6 million tonnes, while consumption remains at 10 million tonnes (CONAB, 2016). Factors such as the incentive to produce, high dependence on meteorological factors and low profitable yields limit the growth of Brazilian wheat production.

Nitrogen fertilisation in wheat

Studies conducted by Viana et al. (2007) show that in order to obtain maximum grain yield, wheat needs to choose the genotype with the best adaptability and stability, as well as high soil fertility. Among the nutrients needed for plant development, nitrogen stands out. According to Scalco et al. (2002), the use of this element in mineral form is indispensable because the amount required by the plant is greater than the concentration available in the soil (SOUZA; FERNANDES, 2006).

Adequate plant nutrition is essential for obtaining high yields from the wheat crop. According to studies carried out by Pettinelli Neto et al. (2002), nitrogen is one of the nutrients most demanded by the cereal. Nitrogen fertilisation tends to be balanced, where excess can cause damage to the environment through the leaching of nitrate into the groundwater, and to the producer through higher production costs. Lack of nitrogen fertilisation has a direct impact on grain yield, which is the most important nutrient in wheat production (LAMOTHE, 1998; SYLVESTER-BRADLEY et al., 2001).

In addition to the quantitative importance, studies carried out by Coelho *et* al. (2001) show that protein content and quality are affected by soil and climate conditions and the availability of the nutrient to the plant. The concentration and quality of protein in wheat is one of the determinants of grain quality, and nitrogen is essential for increasing the quality of the wheat crop (SOARES SOBRINHO, 1999).

However, nitrogen needs to be supplied in the right quantity and at the right time to ensure that the plant is able to express its maximum production potential. Therefore, the phenological stage of the plant is the indicator of when nitrogen fertiliser should be applied. According to the Feeks-Large scale (1954), the wheat cycle *is* divided into five phases and these include other sub-phases. During the tillering period, nitrogen levels must be adequate, as a lack of the nutrient causes significant losses in productivity (BENETT et. al., 2011).

In addition to productivity, the right time to fertilise can reduce the risk of groundwater pollution caused by nitrate accumulation (MAHLER et al., 1994). An alternative used to reduce the effect of nutrient leaching is to fertilise in instalments, as this provides greater assimilation of the nutrient by the plant (MUNDSTOCK, 1999). Fertilising in instalments can also benefit the industrial quality of the grain, as the availability of nitrogen during the period of grain formation and filling contributes to the formation of protein in the grain (ROSA FILHO, 2010).

Poultry litter as an alternative source of nitrogen

The increase in global consumption of poultry meat has fuelled the growth of poultry production in Brazil, and consequently there has been an increase in waste production. One way of avoiding inappropriate exposure to the environment would be to use this organic waste in agriculture as a source of nutrients and organic matter. At the same time as the high cost of chemical fertilisers, the use of organic fertilisers becomes an economic and environmental alternative. Also, as previous studies have shown, the use of this residue improves the physical and chemical properties of the soil, increasing fertility and consequently increasing productivity (SOUTO et al., 2005). According to studies carried out by Felini (2011) when using poultry litter as a fertiliser for soya and maize crops, there was an increase in productivity when doses of up to eight tonnes per hectare were applied.

When comparing the nutrients in poultry litter with mineral fertilisers, it can be seen that since poultry litter has higher levels of nitrogen, phosphorus and potassium, it requires less fertiliser applied to the soil per year in kg, while chemical fertiliser would require 8,090 kg of urea, and N, which *is* equivalent to the amount of urea present in poultry litter, requires only 3,640 kg of poultry litter per year, thus reducing fertiliser costs.

It is known that the various sources used to meet the crop's nutritional needs have a high added value, so the use of poultry litter is an economically and environmentally sustainable alternative. The most practical definition of poultry litter *is the* material used to form the floor of poultry houses, made up of rice straw, grass hay, crushed corn cobs, sawdust, shavings, feathers and leftover feed. Studies are being carried out to understand the use of animal manure for soil fertilisation purposes, with a focus on understanding its composition and its effect on the chemical, physical and biological characteristics of the soil and on plant productivity. The economic aspect of using this manure has also been studied, albeit to a lesser extent, mainly with regard to the logistics of transporting the manure (economically viable distance) and the partial or total replacement of mineral fertiliser with organic fertiliser (PANDOLFO, 2005).

According to Prá (2005), the nitrogen from the solid fraction of the waste is found in the mineralisation process, which includes the amination and ammonification processes, thus transforming this organic nitrogen into mineral form and making it available to the plant. This process is regulated by the use of soil management and, as seen in studies carried out using liquid pig manure, the transformation of nitrogen is efficient, but an extremely slow process (SHERER et al., 1994). Among the determining factors for the mineralisation time of this waste, the type of soil has a direct influence, as its constituents alter the adsorption of these elements which in many cases cause limitations in microbial activity, slowing down the process (SAGGAR et al., 1996).

In order to make use of these resources, it is recommended to follow the Manual of Fertilisation and Liming Recommendations for the states of Rio Grande do Sul and Santa Catarina (CQFS RS/SC, 2004), which sets out the following parameters for recommending organic fertilisation: dry matter content, nutrient release efficiency index (IELN) for organic fertiliser, concentration of nitrogen, phosphorus and potassium. Among the nutrients, nitrogen is essential for wheat, as it increases the growth and development of the crop and has a direct influence on the quality and quantity of the grains.

The use of poultry litter becomes an alternative for the producer, due to the need to dispose of the material, as well as taking advantage of the concentration of nutrients in the composts, which provide the plants with nutrients necessary for development.

CHAPTER 10

Performance of maize hybrids in relation to forage production

Vinícius Jardel Szareski, Ivan Ricardo Carvalho, Maicon Nardino

Brazil has a great aptitude for agribusiness, and in this scenario, beef cattle farming and dairy farming are prominent in the national production chains. Given the various agricultural production scenarios, the southern region has a favourable climate for dairy production, but the supply of quality food is seasonal throughout the year. Silage is therefore economically viable, as it meets the nutritional demands of the diet, proving to be a high quality feed that is readily available to the animals (MITTELMANN et al., 2005).

The quality of the silage produced is the result of factors linked to the growing environment and the ensiling process, soil and climate characteristics, management techniques assigned to the crop, the technological level of the farm, and the characteristics of the maize genotype used (MOREIRA et al., 2001). Thus, maize hybrids used for silage production must have a high dry matter yield per unit area, nutritional value and digestibility (GOMES et al., 2006). The determining characteristics for producing quality silage are attributed to the dry matter fraction, crude protein and energy proportion (PAZIANI et al., 2009). Research has shown that the physiological quality of the seeds, the initial establishment of the plants in the field, the sowing time, fertiliser management, arrangement and

plant population, interspecific competition with weeds, can influence the magnitude and quality of silage in the maize crop (ROSA et al., 2004).

The characters of importance in the production of maize silage are widely influenced by the environment, including the proportion of digestible dry matter, the fraction of starchy carbohydrates, the intrinsic characteristics of the genotype and the interaction between genotypes and environments (MITTELMANN et al., 2005).

In this context, it is crucial to understand which traits are most important when improving maize for silage production, with a view to productivity per unit area, quality of the product obtained, high storage life and digestibility (GOMES et al., 2004). At the same time, it is important to highlight the associations between the traits of interest, identifying the genetic variation of the genotypes involved. With this in mind, the aim of this study was to identify the phenotypic classes and the relative contribution of the traits, identify the linear associations and the genetic variation between the maize hybrids used for silage production.

The work was carried out during the 2012/2013 agricultural season in the municipality of Campos Borges - RS, at latitude 28°52'31"S and longitude 53°01'55"O, with an altitude of 439 metres. The region's soil is classified as Dark Red Latosol (EMBRAPA, 2006), and the climate is characterised by Kõppen as humid subtropical *Cfa* (MORENO, 1961).

The experimental design used was a randomised block design with eight maize hybrids in six replications. The genotypes used were: HT3248, HS1356, HS1380-2, HS1358, HT7, HSM97, HS1380-1 and HT4, totalling 48 experimental units. The experimental units consisted of four five-metre long rows spaced 0.5 metres apart and a population of 75,000 plants per hectare for all the genotypes tested. Sowing took place in the first half of October 2012 and silage was made in the first half of January 2013.

The point at which the plants were harvested for silage was when they showed milky to mealy grains in their ears. The plants were then crushed into particles of between 10 mm and 20 mm and ensiled in PVC mini-silos measuring 3925 cm^3 . Dry sieved sand was added to the base of the mini silos to store the solutes resulting from the silage fermentation process. The silage was compacted to a density of 600 kilos of fresh mass per cubic metre (VELHO et al., 2007), after which the mini-silos were hermetically sealed and stored for 820 days.

Base fertiliser was used with 250 kg ha[1] of NPK in the formulation (10-20-20), and 90 kg ha[1] of nitrogen in the amide form was applied as a top dressing at vegetative stages V4 and V6 (FANCELLI and DOURADO NETO, 2000). Weed and insect pest management was carried out preventively in order to reduce external interference with the results of the experiment. The characters assessed were measured in the useful area of each experimental unit, which consisted of two central rows, where the 0.5 metres at each end were discarded to minimise the effects of the border:

- Plant height (PH), measured from ground level to the last expanded leaf, results in metres (m).

- Ear insertion height (AIE), measured from ground level to the point of insertion of the main ear, results in m.

- Prolificacy (PRO), counting the number of viable ears per plant, results in units.

- Chlorophyll content (CT), obtained using a SPAD-502 chlorophyll meter, in leaves from the middle third of the plant, results in mg m' .

- Leaf area (LA), the length and width of the leaves were measured on five plants per experimental unit, then the values were corrected by the factor 0.75 (MONTGOMERY, 1911), results in cm^2 .

- Green mass yield per hectare (RMV), obtained by crushing the useful area of the experimental unit and measuring the mass of the crushed material, then calculating the ratio between the mass of silage and the number of crushed plants, adjusting for the final plant population used, results in kg ha .[1]

- Dry mass yield per hectare (RMS), the crushed green mass was subjected to drying at 65°C, the results obtained were adjusted for the final plant population used, results in kg ha[1] .

- Percentage of silage dry mass (PERS): 100g of desensilised material was collected. The fractions were placed in an oven at a temperature of 105°C for 24 hours, after which the ratio between the mass of fresh and dry matter was obtained, the results in percentage.

- Silage hydrogenic potential (pH); hemicellulose (HEM); cellulose (CEL); mineral material (MM), were obtained using the technique of Silva & Queiroz (2006), results in percentage.

- Neutral detergent fibre (NDF), the samples were autoclaved with neutral detergent at 110°C for 40 minutes (SENGER et al., 2008), results in percentage.

- Acid detergent fibre (FDA), the residue of the NDF was autoclaved with acid detergent at 110°C for 40 minutes (SENGER et al., 2008), results in percentage.

- Lignin (LIG), measured using the methodology proposed by Robertson & Van Soest (1981), results in per cent.

- Total carbohydrates (TC) and non-fibre carbohydrates (NFC) were obtained using the methodology of Sniffen et al. (1992), results in percentages.

- Crude protein (PTN), determined using the methodology proposed by Nogueira & Souza, (2005), results in per cent.

- Lipids (LIP), determined by the technique presented by Bligh & Dyer (1959), results in per cent.

The analysis of variance revealed significance at 5% probability for the characters mineral material (MM), total crude protein (PTN), hemicellulose (HEM), lipids (LIP), total carbohydrates (CT) and non-fibre carbohydrates (CNF), however, no significant differences were found for the characters plant height (AP), ear insertion height (AIE), prolificacy (PRO), chlorophyll content (TC), hydrogen potential (pH), dry mass percentage (PERS), neutral detergent fibre (NDF), acid detergent fibre (FDA), lignin (LIG), cellulose (CEL), leaf area (AF), green mass yield per hectare (RMV) and dry mass yield per hectare (RMS).

The purpose of frequency analysis is to organise a set of observations, facilitating visualisation, while identifying the classes formed by observations in each character (CRESPO, 2002). The chlorophyll content (CT) character revealed four phenotypic classes, ranging from 14.4 to 16.2 mg m^{2} , so that 37.5% of the maize hybrids were in the 16.2 mg ha^{1} class. Chlorophyll is a group of pigments present in plant tissues and has the ability to transform light energy into chemical energy in the photosynthetic process, being crucial in the transformation of inorganic molecules into organic molecules in the form of assimilates (VOLP et al., 2008), 2009). High chlorophyll levels are a desirable aspect in maize hybrids used for silage, as they can be indicative of the genotype's increased production capacity and quality. The leaf area (LA) character showed four phenotypic classes, ranging from 5550 to 6450 cm, with 37.5% of the maize hybrids evaluated comprising the 6150 cm class^2. The size of the leaf area of a plant suitable for silage production is closely related to the production of green and dry mass per unit area. Research by Cunha (2004) reveals a strong positive relationship between the leaf area index and dry matter production of Tanzania grass.

The plant height (PH) trait showed four phenotypic classes, with a range of 1.52 to 1.67 metres, and the classes were established similarly with 12.5% of the hybrids tested. The increase in plant height is a factor that increases the unit mass of the plant, leaf units and stalks per area, together revealing greater potential for the ensiling process.

The ear insertion height (AIE) revealed four phenotypic classes, ranging from 0.90 to 1.08 m, so that 37.5% of the hybrids were in the 1.08 m class. Ear insertion height is related to the maize plant's balance point, since plants with higher ear insertion are more prone to lodging, which is not a desirable characteristic for both silage and grain production. Maize cutting height is a characteristic that influences the quality of the silage produced, where higher cuts closer to the cob tend to increase the DM of the silage, with a higher proportion of grain in relation to straw, and cuts further away from the cob reduce the DM. In their research, Neumann et al. (2007) concluded that cutting maize at a height of 15.2 m resulted in a 7.1% increase in silage DM production compared to cutting at a height of 38.6 cm from the ground.

The prolificacy (PRO) trait showed three phenotypic classes, ranging from 0.98 to 1.10 m, with 50% of the hybrids in the 1.10 class, indicating that most of the hybrids used are more prolific and may contribute more to the DM of the silage due to the increase in grains and starch fraction in the silage mass. The production of cobs by maize hybrids suitable for silage production is an extremely important aspect, as they reduce the proportion of slowly digestible fibres, increase the protein and starch fraction and provide better plant digestibility (ZAGO, 2002). In this sense, Nússio (1992) defined that the ideal maize plant for silage production should consist of 20 to 23% stalk, 12 to 16% leaves and 64 to 65% cobs.

The character green mass yield per hectare (RMV) revealed four phenotypic classes, ranging from 31250 to 38750 kg ha[-1], with 62.5% of the hybrids in the 38750 kg ha[-1] class. The production of green mass silage is an inherent characteristic of the hybrid used, but it is strongly influenced by the management characteristics. Neumann et al. (2007) found an increase in the production of MV silage from 53744 to 59905 kg ha[-1] when the cutting height was reduced from 38.6 to 15.2 cm. Similarly, the character dry mass yield per hectare (RMS) showed four phenotypic classes, ranging from 8700 to 10500 kg ha[-1], which shows that 37.5 % of the hybrids are in the 8700 and 9900 kg ha[-1] classes. This result falls short of the dry mass yield potential of the hybrids available on the agricultural scene. In studies, Beleze et al. (2003) observed dry mass production of 17.241 ha[1], in a simple hybrid.

The hydrogenic potential (pH) of the silage is distributed in three phenotypic classes, ranging from 3.75 to 3.93, with 3.81 standing out in 50% of the genotypes used. Research by Vieira et al. (2013) with different maize hybrids showed no significant differences in pH between the genotypes; however, the average for this trait was 3.86, in line with the results obtained in this study.

The percentage of dry mass (PERS) character revealed four phenotypic classes, ranging from 27.0 to 29.4%, with 37.5% of the hybrids studied falling into the 27.0% class. The ideal range for the percentage of dry mass in silage varies from 32 to 35%, and the lower the magnitude of this character, the greater the need for consumption by the animal to meet its food requirements (CRUZ et al., 2016). Determining the dry mass of the silage is of fundamental importance in order to base the ideal formulation of the diet, based on the proportion of dry matter in the silage, the daily consumption and the mass of the animal. With regard to mineral material (MM), four phenotypic classes were formed, ranging from 2.7 to 4.5%, with 37.5% of the hybrids in the 2.7% class. The mineral fraction found in maize silage consists mainly of nitrogen, phosphorus, potassium, calcium and magnesium (COELHO & FRANÇA, 1995). According to Underwood & Suttle (1999), minerals play a structural, physiological, catalytic and regulatory role in animal metabolism.

The crude protein content (PTN) revealed five phenotypic classes ranging from 3.0 to 5.4 per cent, with the most prominent class for this character and 50 per cent of the hybrids contained being the 4.8 per cent PTN class. These results are in line with those obtained by Vilela (1985), who defines maize silage as having between 4 and 7% crude protein. In research into the bromatological characterisation of maize silage in the

Brazil, Novinski et al. (2013) points out that the average CP content is 7.1%, although the percentage varies depending on the place of cultivation. Similarly, Vieira et al. (2013) found an average CP percentage of 7.5% in silage from super early maize genotypes.

Neutral detergent fibre (NDF) revealed four phenotypic classes with a range of 50.25 to 54.75%, which corresponds to 37.5% of the hybrids in the 50.25 and 51.75% classes. The NDF content of a roughage is closely related to consumption by the animal, i.e. the higher the NDF content of the ensiled feed, the greater the animal's need to eat in order to satisfy its dietary requirements. Research has defined an ideal fibre fraction of 50% NDF (CRUZ, 1998), so the results obtained in this study are in line with those defined by previous studies. In studies on the bromatological composition of the NDF of maize and sorghum silages, Pereira et al. (2007) showed that using 90% of the panicle for sorghum and 75% of the cobs for maize resulted in lower effective degradation of the silage's NDF.

Acid detergent fibre (FDA) showed four phenotypic classes ranging from 19.5 to 24%, and 37.5% of the hybrids were found to be in the 21.0% class. According to Oliveira et al. (2010), a low percentage of FDA in silage is desirable as it indicates high forage quality and better digestibility conditions. FDA is considered to be a structural polysaccharide, which brings together the slow-digesting fibres. An increase in this fraction tends to show an increase in the difficulty of digesting the silage and a slower rate of passage through the ruminant's digestive system (ALVAREZ et al., 2006).

Lignin (LIG) revealed three phenotypic classes and a range from 0.6 to 1.5 per cent, determining that 37.5 per cent of the genotypes are similarly distributed in the 0.6 and 0.9 per cent classes. Lignin is a polysaccharide present in the plant cell wall. Together with cellulose and hemicellulose, it has a structural function in adult cells (SALMAN, 2010). Lignin makes the plant cell wall more rigid, but hinders the digestibility of forage by the digestive tract of ruminants.

Hemicellulose (HEM) showed four phenotypic classes and a range corresponding to 28.5 to 34.5%, with 37.5% of the hybrids in the 28.5 and 30.0% classes. These results are in line with those obtained by Silva et al. (2005), who showed the effects of inoculation with microorganisms in maize and sorghum silage, and revealed 30.20% of hemicellulose in the silage. The results obtained in this study are in line with those obtained in previous studies.

Cellulose (CEL) revealed four phenotypic classes ranging from 19.5 to 22.5 per cent, with 37.5 per cent of the hybrids in the 20.5 per cent class. Cellulose is the main component of the plant cell wall and is responsible for its rigidity. The results obtained in this study are lower than those obtained by Silva et al. (2005), who obtained 27.45% cellulose in maize silage.

Lipids (LIP) showed the formation of three phenotypic classes and a range from 3.52 to 3.97%, with 50% of the hybrids in the 3.52% class. These results are in line with the levels prescribed by Medeiros et al. (2015), who point out that lipid levels above 6% can have a negative effect on fibre degradation in the rumen; however, despite their limitations, lipids have nutritional properties, being an energy source and essential for ruminant metabolism.

Total carbohydrates (TC) revealed four phenotypic classes and a range from 89.5 to 92.5 per cent, with the 91.5 per cent class standing out with 37.5 per cent of the hybrids. The total carbohydrate fraction represents around 80% of the dry mass of forage and includes the protein, starch, lipid and fibre fractions (VAN SOEST, 1994). Non-fibre carbohydrates (NFC) showed three phenotypic classes ranging from 32 to 38%, with 50% of the hybrids in the 38% NFC class. This result corroborates the results obtained by Cabral et al. (2004), which revealed a percentage of 37.05% NFC in maize silage. Non-fibrous carbohydrates comprise the fractions of starch, sugars, pectin and volatile fatty acids present in the feed and are characterised by being highly energetic and highly digestible in the rumen

(CABRAL et al., 2004). The NFC fraction present in maize silage is made up predominantly of the starch fraction (71.3%), and lipid fraction with 28.7% (NRC, 2001).

The purpose of *Pearson*'s linear correlation analysis is to identify the trend of linear association between characters. It is expressed by correlation coefficients that range in magnitude from -1 to 1 with positive or negative meanings. These coefficients are classified as null (r=0.00), low or weak (r=0.00 to r=0.30), intermediate or medium (r=0.30 to r=0.60); high or strong (r=0.60 to r=1.00), according to Carvalho et al. (2004). *Pearson*'s linear correlation was carried out for the eight maize hybrids and totalled N=48, where the linear associations were carried out for the 19 characters: AP, AIE, PRO, TC, AF, RMV, RMS, PH, PERS, MM, PTN, FDN, FDA, LIG, HEM, CEL, LIP, CT and CNF. A total of 171 associations were obtained, of which only 52 were significant using the *t-test* at 5% probability.

Plant height (PH) showed a high and positive correlation with EIA (r=0.78), an intermediate and positive correlation with TC (r=0.58), AF (r=0.31), RMV (r=0.52), RMS (r=0.47), pH (r=O.35), FDN (r=0.49), FDA (r=0.40), HEM (r=0.32), CEL (r=0.43), but an intermediate and negative trend with CNF (r=-0.47). Thus, plants with greater stature tend to increase their leaf area and chlorophyll content, thereby increasing their photosynthetic efficiency and accumulation of assimilates, resulting in higher yields of green and dry mass per unit area, increasing structural fibrous carbohydrates and reducing non-fibrous carbohydrates in silage. Cancellier et al. (2011) and Pazianini et al. (2009) revealed a strong positive correlation between PA and EIA, as well as between PA and RMV. In maize, taller plants can increase green mass yield and thus boost silage production (MELO et al., 2004). However, taller plants can have problems with lodging, which are mitigated in maize for silage production, as the plants spend less time in the field exposed to adverse weather conditions (CANCELLIER et al., 2011).

Ear insertion height (AIE) showed an intermediate and positive correlation with TC (r=0.46), AF (r=0.33), RMV (r=0.47), RMS (r=0.46), but a weak and negative trend with CNF (r=-0.48). Therefore, plants with taller ears tend to increase the production of green and dry mass, but may reduce the content of non-fibrous carbohydrates in the silage. Research by Cancellier et al. (2011) revealed a positive correlation between AIE and RMV in open-pollinated maize varieties.

Prolificacy (PRO) and leaf area (AF) showed an intermediate and positive correlation with HEM (r=0.30). This indicates a tendency for more prolific plants with greater leaf area to increase the hemicellulose content in silage. Hemicellulose is associated in the plant with lignin and structural polysaccharides. Older plants tend to increase the hemicellulose fraction in proportion to the amount of leaf area and stalk mass present in the maize plant (JÚNIOR et al., 2007).

Chlorophyll content (CT) showed an intermediate and positive correlation with RMV (r=0.58), RMS (r=0.51), FDN (r=0.37), FDA (r=0.32), CEL (r=0.34), a weak and positive correlation with AF (r=0.29), and an intermediate and negative correlation with CNF (r=-0.37). This indicates that plants with a higher chlorophyll content increase their photosynthetic capacity and increase the yield of green and dry mass per unit area, but reduce the fraction of non-fibrous carbohydrates. Non-fibrous carbohydrates differ from fibrous carbohydrates due to the presence of pectins, hemicellulose and cellulose (CABRAL et al., 2002). The non-fibrous carbohydrates present in maize silage consist of starch, proteins and lipids (CABRAL et al., 2004).

Green mass yield per hectare (RMV) showed a strong positive correlation with RMS (r=0.90), and a weak positive correlation with LIP (r=0.30). This result shows a positive association between the green and dry mass yield of maize silage and the lipid content of the silage. The percentage of dry mass (PERS) showed an intermediate and positive trend with NFC (r=0.54), intermediate and negative with NDF (r=- 0.47), HEM (r=-0.46), CEL (r=-0.35), and weak and negative with MM (r=- 0.29). Thus, increasing the percentage of dry mass in silage tends to increase the proportion of fibrous and non-fibrous carbohydrates, as well as the inorganic fraction.

Mineral material (MM) showed an intermediate and positive correlation with HEM (r=0.31), and an intermediate and negative correlation with PTN (r=-0.42),

LIG (r=-0.36), CT (r=-0.55) and CTN (r=-0.51). This indicates that an increase in the mineral fraction tends to reduce the organic fraction of silage, while increasing hemicellulose. Crude protein (PTN) showed an intermediate and positive correlation with LIG (r=0.43), and an intermediate and negative correlation with FDN (r=-0.35), HEM (r=-0.36) and CT (r=-0.32). The high protein concentration results in an increase in the lignin content, reducing the neutral detergent fibres, hemicellulose and total carbohydrates of maize silage.

Neutral detergent fibre (NDF) showed a strong positive correlation with FDA (r=0.75), HEM (r=0.72), CEL (r=0.74) and a strong negative correlation with NFC (r=-0.95). Fibre content in plants is of great importance for silage quality (GRALAK et al., 2014). The NDF fraction of the feed is decisive in defining the amount of feed required to meet the animal's dietary needs, while FDA and lignin reduce the digestibility of the feed (MORAES et al., 2013).

Acid detergent fibre (FDA) showed a strong positive correlation with CEL (r=0.96), a strong negative correlation with CNF (r=-0.72) and an intermediate negative correlation with CT (r=-0.35). The percentage of cellulose is increased in plants with higher FDA (PIRES et al., 2010).

Hemicellulose (HEM) and cellulose (r=-0.68) showed a strong negative correlation with NFC (r=-0.71). This indicates that plants with a higher content of hemicellulose and cellulose have lower non-fibrous carbohydrates because these components are intrinsic to the plant cell wall and are common in the stalk, while non-fibrous carbohydrates are abundantly present in the grain (CABRAL et al., 2002).

The analysis of canonical variables is a tool that makes it possible to identify the genetic variation involved between the maize hybrids studied, this technique makes it possible to highlight the arrangement of genotypes through a three-dimensional plane (MIRANDA et al., 2003).In view of this, the analysis was used for the 19 characters measured, in this way, only three canonical variables were necessary to represent 78.95% of the total genetic variation, where VC1 represents 38.23%; VC2:36.61% and VC3:9.64%.

The analysis of canonical variables revealed the formation of five groups to differentiate the eight maize hybrids: group I made up of genotypes HT3248, HS1380-2 and HT4, group II made up of HSM97 and HS 1380-1, group III represented by hybrid HS1356, group IV by HS1358 and group V by genotype HT7.

The relative contribution of the traits defined by Singh's method (1981) shows that the determining traits in distinguishing maize genotypes for silage were neutral detergent fibre, non-fibrous carbohydrates and hemicellulose, while the traits plant height and prolificacy made the smallest contribution to distinguishing the genotypes.

The frequency distributions indicate that the characters with the highest number of phenotypic classes were total crude protein and hemicellulose.

Cellulose is highly related to the fraction of structural carbohydrates that are difficult to digest. The percentage of cellulose and hemicellulose is inversely proportional to the non-fibrous carbohydrates in maize silage.

The traits neutral detergent fibre, non-fibre carbohydrates and hemicellulose are those that contribute most to the discrimination of maize genotypes. The contribution of descriptive, univariate and multivariate analyses are techniques that can be used successfully to discriminate between maize genotypes and characters of importance for the production of maize silage.

CHAPTER 11

Fertilisation and maize seed production

Ivan Ricardo Carvalho, Vinícius Jardel Szareski, Maicon Nardino

In maize growing, inadequate nitrogen (N) management is considered one of the main factors limiting grain yields. N is important in plant metabolism, as it participates in the biosynthesis of proteins and chlorophylls, and is needed from the initial phenological stages of plant development (BASSO and CERETTA, 2000), and according to Sangoi et al. (2008) it participates in numerous metabolic routes, important in plant biochemistry (ANDRADE et al., 2003), Results of research carried out by Coelho (2004) and Lemaire & Gastal (1997) show that with adequate levels of other nutrients in the soil, nitrogen provides the greatest increases in maize productivity.

Due to the various transformations that N undergoes in the soil, this nutrient is considered a dynamic element with complex reactions, which generates controversy and discussion regarding its source and the timing of its application to maize. In the soil-plant system, its dynamics are influenced by many characteristics in the cultivation systems (conventional or no-till), the forms of management, soil and climate conditions (SANTOS et al., 2010) and the type of fertiliser. Currently, urea and ammonium sulphate are the main nitrogen sources used to grow maize. According to Alva et al. (2006) both in the soil are subject to losses through leaching, surface run-off, ammonia volatilisation and immobilisation in the microbial biomass.

Nitrogen is released to plants depending on the nitrogen source. Urea $[CO(NH_2)2]$, which is in the amide form, is rapidly hydrolysed in the soil, producing ammonium $[NH4^+]$, making it possible for it to be retained by the negative charges of soil colloids and readily available to plants (NOVAIS etal., 2007). However, some authors have observed that $NH4^+$ from urea tends to be nitrified more quickly than $NH4^+$ from ammonium sulphate, due to the increase in the pH of the medium during hydrolysis (MCINNES & FILLERY, 1989; SILVA & VALLE, 2000).

Ammonium sulphate $[(NH4)_2 SO_4]$ when applied to the soil in the dissolution process directly produces ammonium ions, which can be oxidised to nitrate in the nitrification process (SILVA & VALLE, 2000). On the other hand, depending on climatic conditions, ammonium sulphate can delay the release of N due to the fact that it is protected by organic sulphur. According to Queiroz et al. (2011) the coating of organic and inorganic substances allows the nutrient to be released gradually, while (TRENKEL, 2010) the release can be delayed due to the slow hydrolysis of water-soluble compounds.

One factor that can influence the nitrification of nitrogen from ammonium sulphate and urea is the concentration of calcium in the soil. Work carried out by Aurora et al. (1986) observed that ammonium from urea nitrified more quickly than that from ammonium sulphate in the absence of calcium carbonate.

Nitrogen assimilation can vary between hybrids. According to Machado (1992), genetic differences can influence the activity and assimilation of nitrogen by maize plants. Studies carried out by this author conclude that hybrids that incorporate the ammonium ion into amino acids through the enzyme glutamine synthetase, have better efficiency in utilising nitrogen when compared to enzymes present in other hybrids. This result correlates with that found by Femandes et al. (2005) that nitrogen doses express different efficiencies in maize hybrids. Other factors that may affect the response to nitrogen fertilisation are genetically modified hybrids, which accounted for 42% of the maize seeds sold in the 2010 off-season. This is reflected in the growth in maize productivity with the use of hybrid seeds with greater genetic potential and greater use of fertilisers (CRUZ et al., 2010).

Currently, the nitrogen fertiliser recommendation for maize in Rio Grande do Sul is between 20 and 30 kg of N at sowing, and the rest in top dressing between phenological stages V4 to V6 (CQFS- RS/SC, (2004), but Escosteguy et al. (1997)

state that top dressing can vary between four and eight fully expanded leaves. An extremely important factor in the management of N in maize crops is the amount of nitrogen applied, because, according to Kappes et al. (2013), maize yields increase as the dose of N increases. In this sense, there are concerns about this nutrient which are linked to application practices, where losses in certain climatic conditions are evident, because numerous chemical reactions influence its release and absorption by plants. Due to the transformations that N is subject to in the soil, studies related to sources and timing of nitrogen in maize have generated controversy and much discussion (MEIRA et al., 2009).

Soratto et al. (2012), when studying nitrogen sources and instalments in maize, found no difference in instalments, but observed an increase in productivity between sources. Schiavinatti et al. (2011), on the other hand, found negative effects on productivity when urea was applied in a single application at the eight-leaf stage, but there were no differences between nitrogen sources.

In the literature, the results are conflicting for the timing of nitrogen application to maize, which is why the hypothesis behind this work is that the instalment and source of nitrogen can affect some plant parameters and final grain yield, and that different responses can occur between genetically modified hybrids. In addition, the current recommendation is based on row spacings of 80-90 cm. However, reduced spacings are currently used, which can directly influence the use of nitrogen fertiliser by the maize crop. For this reason, parceling out the fertiliser can improve the efficiency of nitrogen assimilation by plants and bring significant gains in grain yield, requiring techniques to reduce losses, define the best source and associate the best time for nitrogen application. The aim of this study was therefore to assess the effect of nitrogen instalments on some plant parameters and maize yields of two hybrids, using urea and ammonium sulphate.

Nitrogen to maize ratio

Among its functions, nitrogen is important in plant metabolism. According to

Sangoi et al. (2008), this element participates in various metabolic routes that are extremely important to plants, and according to Vieira et al. (1995) in protein synthesis. It is a constituent of biomolecules such as ATP, NADH, NADPH, storage proteins, nucleic acids and enzymes (HARPER, 1994), a constituent of cytochrome molecules and chlorophyll (BULL, 1993). This shows that nitrogen is directly correlated with plant development and productivity.

Most cultivated soils allow plants to grow without the addition of nutrients in mineral form, but when looking for higher production levels, it is necessary to apply nutrients in mineral form. Among the main crops of agronomic interest, maize is highly dependent on nutrients, especially nitrogen (CANCELLIER 2011). Under favourable climatic conditions, for high yields maize uses amounts of nitrogen in excess of 150 kg[l] , requiring mineral supplementation with nitrogen fertiliser sources (AMADO 2002; SCALCO et al. 2002), Studies carried out by Taiz & Zeiger (2009) show that maize is directly dependent on the use of nutrients in mineral form, where large quantities have a positive effect on grain yield. Work carried out by Silva et al. (2005) shows that this nutrient is limiting for the establishment of the crop, and according to Calonego (2012) the efficiency of absorption and translocation of N into the grain is an extremely important factor, as it directly influences productivity.

Research reports numerous results regarding the importance of nitrogen for maize cultivation. According to Basiet al. (2011), N has an influence on the quality of maize silage, stating that the quality of the grains is positively affected when nitrogen is used, as a plant well nourished with nitrogen produces silage with a higher nutritional value. Ferreira et al. (2001), when studying agronomic characteristics with the use of nitrogen, concluded that nitrogen fertilisation improved grain quality, increasing protein and mineral nutrient content, positively affecting the number of ears per plant, ear mass, and the mass of a thousand seeds, increasing according to nitrogen doses.Another important factor in determining nitrogen fertilisation in maize is the difference in the use and assimilation of N between hybrids (NUNES et al., 2013), 2013).

Nitrogen in the soil-plant system

N in the soil is associated with soil organic matter (SO), so much so that in current nitrogen fertiliser recommendations this is one of the criteria for defining the amount to be applied. More than 95% of the nitrogen in the soil is in organic form. This organic form is not assimilated by the plant and has to go through a mineralisation process, i.e. the transformation of organic nitrogen into mineral nitrogen called amination and ammonification (RANGEL & SILVA, 2007). According to Mary et al. (1996) the time and amount of nitrogen mineralised can vary depending on the species, amount of organic residue, temperature, aeration and humidity. Therefore, the total N reserve present in the soil does not always represent the availability of the nutrient for plants (AMADO et al., 2002).

With the adoption of the direct sowing system, nitrogen fertilisation has become even more important, especially when the preceding crop is grass, due to the large amount absorbed and the lower rate of decomposition. According to Ceretta et al. (2002) the decomposition of plant residues is directly associated with the carbon/nitrogen ratio (C/N) of the plant tissue, which will reflect on the decomposition of the residue and the immobilisation of N, influencing the availability of nitrogen.

Nitrogen stands out among the essential nutrients for plants, as it undergoes a series of transformations in the soil-plant system (STEVENSO, 1982), which are carried out by microorganisms (VICTÓRIA et al., 1992), making its dynamics in the soil extremely complex. As this nutrient undergoes various reactions in the soil, depending on the environmental conditions, it can be subject to losses, especially when applied in large quantities, causing damage to the environment. According to Silva (2005) there are countless ways in which nitrogen leaves the soil, including gaseous losses through ammonia volatilisation, denitrification of NO_3 ' to N_2 , leaching losses, erosion and nutrient extraction by plants.

Volatilisation is the loss of N in gaseous form, which occurs with both urea and ammonium sulphate. Factors such as temperature, soil pH and the method of application are the main factors affecting loss by volatilisation. Another form of loss is denitrification. In this process, in the absence of oxygen, microorganisms can fulfil the need for oxygen by using the nitrate ion (NO_3 '), resulting in nitrite (NO_2 '). Losses also occur through leaching, where nitrogen can be lost through movement at depth, being carried away by water in the form of (NO_3), (NH_4 ') or humified organic compounds (REICHARDT & TIMM, 2012). According to Coelho (2003), the leaching of NO_3 " occurs due to its high mobility.

There are two forms in which nitrogen is present in the soil and available to plants, in the nitric form (N-NO_3 ') and in the ammoniacal form (N-$NH4^+$) (ARAÚJO et al., 2012), which is absorbed by the roots and translocated to leaves and stems (REICHARDT & TIMM, 2012).Various studies show that the utilisation of nitrogen when applied in unsuitable environmental conditions rarely exceeds 50% (SCIVITTARRO et al., 2000).

As soon as the maize emerges, the roots begin to absorb nitrogen. According to Ta & Weiland (1992), the N absorbed during initial development has a structural function in the plant, and little is stored and translocated.

Work carried out by Ferreira et al. (2001) reported that N quantified in the leaves at 25 DAE (Days after emergence) was not correlated with grain yield, but at 45 and 63 DAE it was positively correlated with grain yield. França et al. (2011) observed that nitrogen absorption was ascending, reaching a maximum value at 74 DAE in milky grains, and from then on there was a reduction in N in the biomass. This result differs from that found by Fomasieri (1992) where the greatest demand for N occurred two to three weeks before flowering and the reduction was at 82 DAE, due to the mobilisation of N into the grains, leaching of N by leaves, leaf fall and senescent stalks.

Like nitrogen, the nutrients accumulated in the plant parts are translocated to the grains during the grain-filling period, with a positive correlation with the chemical composition of the leaves and the accumulation of protein and nutrients in the maize grains (FERREIRA et al., 2001).

Morphological components and maize productivity

Identifying the yield component that makes the greatest contribution to maize productivity is an important tool that helps to define the critical period for crop development, enabling management practices to be adopted with the aim of identifying the moment when the main grain yield component will be defined (BALBITON et al., 2005).

The number of grains produced per area is the yield component that most interferes with grain yield, which is modified by the plant's ability to produce and distribute photoassimilates to meet the demand after the spike period (SANGOI et al., 2005). This is shown by the results found by Souza et al. (2014) who, when studying the relationship between maize plant traits and final grain yield, concluded that the variable ear height shows a positive correlation with grain yield, making it an appropriate trait for indirect selection for yield. They also concluded that the variable stalk diameter had a direct negative effect on grain yield, where plants with larger stalks had lower yields.

A result found by Balbiton et al. (2005) when assessing the variation in grain yield between open-pollinated varieties (OPV) of maize and their yield components, concluded that there is variability in grain yield and yield components between maize varieties, since in this study the number of grains per row was the yield component that contributed most to productivity. Another result that demonstrates the importance of the correlation between grain yield and yield components is that presented by Sangoi et al. (2011), who assessed the influence of N on the morphogenesis and tillering processes of maize and concluded that tillering does not directly contribute to maize grain yield.

Nitrogen in the direct seeding system

Since 1990, the direct seeding system has been intensified. This system is based on the premise of not turning the soil and keeping the straw on the surface. As a result, there has been a change in the way nitrogen is managed in the maize crop, seeking to develop strategies to improve nitrogen efficiency. According to Basso & Ceretta (2000) in direct sowing we should increase the availability of nitrogen in mineral form, especially when the preceding crop is grass.

According to Amado (2002), this crop has the capacity to absorb and accumulate nitrogen in the aerial part, but a study carried out by the same author evaluating the decomposition of oats found that only 20% of the N present in the plants was released in the first four weeks. This is correlated with the C/N ratio, which has benefits such as keeping plant residues on the soil for longer, but results in lower nitrogen availability for the successor crop (COLLIER et al., 2011).

On the other hand, leguminous plants have a lower C/N ratio, making more nitrogen available to the successor crop, and according to Santos et al. (2010) this is because legumes fix nitrogen and add it to the system. According to Amado (2002), they have the capacity to recycle nutrients and synchronise their slow release, increasing the availability of nitrogen in the soil in the medium and long term. Aita et al. (2001) studied the release of nitrogen in winter legumes and crucifers and concluded that 70% of the nitrogen in the phytomass was released in four weeks.

With the aim of improving nitrogen efficiency in a direct sowing system, several researchers have adapted nitrogen management in the maize crop. Ceretta et al. (2002) evaluated the possibility of partially or totally transferring the nitrogen fertiliser that would be applied as a top dressing to maize, to the tillering of the oats or the pre-sowing of the maize, and concluded that this is not the best strategy. These results corroborate those obtained by Basso & Ceretta (2000) when they studied the application of N to maize in succession to different cover crops, and concluded that it is safer to apply N at sowing and in top dressing rather than pre-sowing.

Increased N availability for maize can be achieved by using ground cover plants as a source of nitrogen (AITA et al., 2001; AITA & GIACOMINI, 2003). Work carried out by Heinrichs et al. (2001) revealed that growing maize in succession to the black oat + vetch consortium substitutes for urea in maize, and also concluded that the yield of maize in the 90% vetch + 10% oat consortium was equivalent to that obtained in fallow with 75 kg[1] of urea.

Nitrogen fertiliser recommendations for maize in Rio Grande do Sul are based on the organic matter content of the soil, predecessor crop and yield expectations (CQFS-RS/SC, 2004). It is recommended in the direct sowing system, when the predecessor crop is grass, to apply 20 to 30 kg of N ha[1] in the sowing line, and when it is leguminous, 10 to 15 kg, and the rest of the nitrogen in top dressing between four and six leaves.

Currently, nitrogen fertiliser management strategies have been adopted with the aim of improving the efficiency and use of nitrogen by maize plants and increasing productivity (CRUZ et al., 2008). The best use of nitrogen is directly linked to the time of application, as very early and/or late applications can be poorly used by the plants (SILVA et al., 2005). According to Mengel & Barber (1974), the time of application of nitrogen to maize crops directly influences how it is used by the plants. This may be linked to the results obtained by Silva et al. (2005) who, when studying urea doses and their instalmenting, found a higher IL (Profitability Index) when half was applied at sowing and the rest between four and six leaves. However, results obtained by Raganin et al. (2010) and Arf et al. (2007) found no difference between the times of nitrogen application.

Nitrogen can behave differently between maize hybrids, and can therefore influence the timing of nitrogen fertilisation. This is confirmed by the results obtained by Femandes et al. (2005) where the influence of maize hybrids revealed similarities in nitrogen efficiency between two of the three hybrids. On the other hand, Júnior et al. (2008), when assessing nitrogen efficiency in two maize hybrids, found differences in nitrogen utilisation efficiency.

In general, the lack of nitrogen in the early phenological stages reduces crop yields. According to Pereira et al. (2013), this is due to the definition of the fourth to sixth leaf in this period, flower differentiation and the end of the leaf differentiation phase. And according to Magalhães et al. (2003) because in the V5 stage we have the initiation of leaves and ears, in V8 the number of rows and grains is defined, and the phenological stages are defined with the visible formation of the collar at the insertion of the leaf sheath with the stalk, divided into the vegetative and reproductive phases.

Nitrogen sources for maize cultivation

There are many sources of nitrogen fertiliser available, but in Brazil most nitrogen fertilisation in mineral form is the use of urea and ammonium sulphate (MEIRA et al., 2009).

Urea [$CO(NH_2)2$] has a high concentration of nitrogen, high solubility and lower cost/unit of nutrient, but higher volatilisation (CALONEGO, 2012). The greater volatilisation occurs because when urea comes into contact with the urease enzyme present in plant waste, it undergoes hydrolysis, producing ammonium carbonate [(NH^CCM, causing ammonia gas (NH_3) to be emitted into the atmosphere (COSTA 2001). Urea applied under suitable conditions undergoes hydrolysis resulting in the utilisation of an H ion$^+$, consequently promoting an increase in pH (SINGH & NYE, 1984; KIEHL, 1989). The increase in soil pH in the direct sowing system caused by the application of lime to the surface layers of the soil is correlated with nitrogen losses. According to Tasca et al. (2011), the volatilisation of NH_3 from urea increases in parallel with the dose of nitrogen and soil pH.

Ammonium sulphate ($(NH4)_2 SO_4$) as well as being a source of nitrogen provides sulphur, which is advantageous when soils are deficient (MALAVOLTA, 1989). The ammoniacal form attaches itself to the soil's clay particles (ZHOU et al., 2011), and a pH of less than seven does not show losses due to volatilisation of ammoniacal nitrogen ($N-NH_3$) (FILHO et al., 2010).

Working with different nitrogen sources (CARMO et al., 2012) found no significant difference in the components evaluated between nitrogen sources. Sangoi (2009) when assessing the effect on germination and initial development of maize plants using different doses of urea and ammonium sulphate observed that urea was detrimental to germination and initial growth when compared to ammonium sulphate. Cabezas & Souza (2008) reported lower maize yields when using urea compared to ammonium sulphate. Eurides et al. (2008) reported lower costs with urea compared to other nitrogen sources.

In general, nitrogen is found in the soil solution in the form of nitrate and ammonium, but physiologically, plants are more responsive to nitric than ammoniacal nutrition (DUETE et al., 2009). Cancellier et al. (2011), when evaluating the efficiency of nitrogen use in maize, found differences in the responses of different genetic materials.

CHAPTER 12

Selection of maize genotypes for seed production

Maicon Nardino, Ivan Ricardo Carvalho, Vinícius Jardel Szareski

Every harvest, new hybrids are produced in company breeding programmes, which undergo preliminary evaluations. One of the setbacks often faced by breeders when selecting and recommending genotypes is quantifying the magnitude of the genotype x environment interaction, emphasising that it is up to the breeder to adopt strategies that reduce or take advantage of these effects (CRUZ; REGAZZI, 2001).

During preliminary evaluations, there are several candidate genotypes to be included in the select group, requiring the breeder to use appropriate tools to ensure efficient selection. Adopting quantitative genetics methods with coherent statistical methods can result in more accurate estimates of genetic and residual variance components, thereby improving predictions of the genotypic value of individuals.

The use of mixed models or individual BLUP (Resende et al., 1996) introduced modifications to the estimation of variance components and genetic parameters, where covariances were estimated and interpreted in terms of their mathematical expectation, thus generating variance components. The REML/BLUP procedures developed later allow the variance components to be estimated directly from the data set, as well as the variances of the random effects.

The restricted maximum likelihood (REML) method proposed by (PATTERSON; THOMPSON, 1971) has become the standard method for estimating variance components and genetic parameters, especially for trials with unbalanced data (FILHO; DE RESENDE, 2000). The preference for this method stems from its statistical properties being superior to the least squares estimator methods (SEARLE; CASELLA; MCCULLOCH, 1992).

The mixed models technique has been used by many researchers in the field of genetic improvement in crops such as eucalyptus ((DE RESENDE et al., 1996));

((ROSADO et al., 2012)), coffee (RESENDE et al., 2001), potato (DE SOUZA et al., 2005), sugar cane (ZENI-NETO et al., 2008), maize (ARNHOLD et al., 2012), (DOVALE et al., 2013), rice (BORGES et al., 2009); (DOVALE et al., 2013). Thus, the standard analytical procedure recommended for studies involving quantitative genetics is REML/BLUP, with estimates of variance components by restricted maximum likelihood (REML) and unbiased linear prediction (BLUP) (RESENDE et al., 2001).

In this context, the adoption of models that estimate genetic parameters and correctly predict genotypic values is important to increase the efficiency of breeders in making decisions during the selection process for pre-commercial hybrids, especially when there is a select group of superior hybrids and they are looking for materials with superior performance within this group. The aim of this study was to estimate the variance components via REML and predict the genotypic values of pre-commercial maize hybrids via BLUP.

The trial was conducted in the municipality of Frederico Westphalen. The pre-commercial hybrids used in the trials come from the breeding programme of the company KSP Sementes Ltda, located in the municipality of Pato Branco-PR.

A randomised block design with three replications was used. The experimental plots were made up of two five-metre long rows, with 42 plants per plot, spaced 0.70 m apart, corresponding to a sowing density of 60,000 plants ha^{-1}. Cultivation was carried out in accordance with the technical recommendations for growing maize. The nails were opened with a mechanised seeder and chemical fertiliser was applied at the same time.

The trial used 24 pre-commercial simple hybrids (KSP Sementes Ltda.). The hybrids were sown according to agro-climatic zoning.

The following agronomic characteristics were assessed for each experimental plot:

- Ear insertion height (AE): refers to the height from the ground to the insertion of the first ear, given in cm;

- Plant height (PH): refers to the height from the ground to the insertion of the node of the last leaf of the plant, given in cm;

- Spike diameter (SD): refers to the average diameter of the central part of three spikes, taken as the average for the three spikes, given in mm;

- Average cob length (CE): refers to the average length of three cobs, given in centimetres;

- Average number of rows (NF): refers to the count of the number of rows of three ears, given in units;

- Number of grains per row (NGF): refers to the count of the average number of rows of three ears, given in units;

- Grain yield in bags per hectare (GY): refers to the total grain yield in bags per hectare, corrected to 13% moisture and for a stand of 42 plants in the two rows;

- Hundred-grain mass (GKM): refers to the average mass of 100 grains counted on three ears of corn, given in g.

The *deviance* analysis revealed significant differences using the likelihood ratio test (LTR) at a 5% probability of error using the Chi square test, considering 1 degree of freedom for all the characters analysed in the experiment. In this sense, the *deviance* analysis showed significant genetic differences between the 24 pre-commercial hybrids evaluated in this study. The estimates of the variance components and genetic parameters.

With regard to the variance component estimates, it can be seen that for plant height 53 per cent of the phenotypic variance is of an environmental nature and 47 per cent is of a genetic nature. For the ear height variable, the genetic component contributes 61 per cent of the phenotypic variance and 39 per cent is environmental. Ear diameter shows a greater share of the environmental component, 51.3 per cent, compared to 48.7 per cent, which is genetic. The ear length character shows a more pronounced effect of the environment on the phenotype, with 68.7% compared to 31.3% of the genetic variance component. The number of rows variable reveals that almost 50% of the phenotypic variance is genetic and 50% is environmental, and the same behaviour is revealed by the number of grains per row character. With regard to grain yield, which is considered a trait influenced by many genes as well as the environment, 35% of the variance is genetic and 65% is non-genetic. For the mass of a hundred grains, the partitioning of the components of phenotypic variance into genetic and non-genetic reveals very similar proportions.

The estimates of selective accuracy for the characters evaluated, which is related to greater accuracy in assessing the predicted genetic value. It should be noted that in this study the values were classified as

from moderate 0.71 to strong 0.90 (CARGNELUTTI FILHO; STORCK, 2007), which indicates that the predicted values were close to the true genetic values of the hybrids. These results are close to those obtained in a study by (NARDINO et al., 2016), which evaluated different simple maize hybrids grown in different locations in the southern region of Brazil.

The broad-sense heritability of individual plots (h^2 g) revealed estimates of low (0.26) to medium magnitude (0.60). This parameter estimate is obtained from the ratio between the $l^2{}_{Cj}{}^2{}_F$ and their respective standard deviations, which indicates that the effects of the environment for some characters, especially those of low magnitude, were more pronounced. The average heritability (h^2 ml) showed moderate (0.51) to high (0.82) estimates. The magnitudes of the heritability parameter in this study indicate that the experimental arrangement was adequate to control the effects of the environment and that the predicted genotypic values were adequately estimated.

The genetic coefficient of variation (CVgi) indicates how much the genotypic fraction represents in the total variation, while the residual coefficient of variation (CVe) reveals how much the environmental fraction is contributing to the total variation. The magnitudes of these two coefficients are reflected in the selective accuracy. In this sense, the higher the selective accuracy, the smaller the absolute deviations between the parametric genetic values and the estimated or predicted genetic values. In the same sense, it can be inferred that close estimates between the coefficients, with a ratio close to 1, show that the experimental quality was adequate, in line with the accuracy estimates highlighted above. According to (VENCOVSKY, 1987), when a ratio between CV_{gi} and the coefficient of residual variation CV_e of 1 or more is revealed, there is a favourable situation for obtaining selection gains.

The estimate of the variance of the prediction error of the predicted genotypic values (PEV), assuming no loss of plots, was low for most of the traits in this study, although the greatest variance was observed for the grain yield trait, although given the nature of this trait it is expected that more pronounced effects of variation will act on this trait.

The estimate of the standard deviation of the predicted genotypic value, assuming no loss of plots, shows low deviations for most of the traits, but for grain yield the deviations are more marked, but this trait has a quantitative inheritance, and non-hereditary effects are more active on this component.

The results of the predicted genotypic means (BLUP) for the characters plant height (AP), ear insertion height (AE), ear diameter (DE) and ear length (CE) are presented.

With regard to the AP character, the KSP18, KSP4 and KSP21 hybrids showed greater plant height than the hybrids ranked after the 11th hybrid (KSP12). However, plant breeding has been working on reducing plant height in current hybrids, so hybrids KSP1, KSP5 and KSP15 are more promising from the point of view of maize breeding. Reducing plant height is an objective for selection, because according to (DE JESUS FREITAS et al., 2013) tall plants are more susceptible to breaking and lodging, especially in regions with a high incidence of wind.

For the ear insertion height character, hybrids KSP19 and KSP17 showed the highest genotypic average by BLUP prediction. AE is carefully evaluated by maize breeders, who are currently looking for plants with a lower centre of gravity, both because of the source-drain relationship and to avoid losses due to lodging. Thus, hybrids KSP9, KSP5 and KSP7 showed the lowest genotypic mean among the set of hybrids evaluated.

When analysing the predicted genotypic value for ear diameter, hybrids KSP20, KSP17, KSP18, KSP8 and KSP22 showed a higher genotypic value than the hybrids ranked from 13th onwards (KSP21). The hybrids with the lowest genotypic values were KSP11 and KSP10. Larger ear diameters are often sought after in breeding, as they allow for ears with a greater number of grain rows. However, it is important to observe grain depth and ear length, so that there is a balance between the three traits and thus high productivity levels.

In terms of ear length, hybrids KSP5, KSP22, KSP7 and KSP13 are superior to the genotypes ranked from 17th onwards (KSP2). It can be inferred that the hybrids with the longest spike lengths generally do not have the longest spike lengths, where longer spikes make it possible to form rows with a greater number of grains, which in turn is related to grain yield.

The characters number of rows (NF), number of grains per row (NGF), grain yield (RG) and hundred grain mass (MCG) can be seen in Table 4. For the variable number of rows, the genotypic BLUP values of hybrids KSP17, KSP19 were higher than the hybrids ranked from the 8th° hybrid (KSP8), considering a confidence interval for separating the hybrids. Maize hybrids with a higher number of rows per ear can provide a higher number of grains per ear, but a balance between NF, NGF and grain mass is necessary to achieve a higher grain yield per hectare.

With regard to the number of grains per row, the hybrids KSP5, KSP4, KSP23 and KSP10 were superior, considering a confidence interval, to the hybrids ranked after the 7th hybrid (KSP16).° It can be inferred that the hybrids with the highest NF were not the same ones that showed the highest number of NGF, indicating that ears with a larger diameter and NF have a shorter length and consequently NGF.

Among the variables studied by genetic improvement, grain yield is the target variable for all plant breeders, as it is directly associated with economic return. Adopting methods such as mixed models to evaluate pre-commercial hybrids that are already considered a select group by plant breeders is of fundamental importance in order to maximise the accuracy of predicting the real genotypic value of individuals, removing biased information from non-genetic effects. Once the superior hybrids within the pre-commercial group have been defined, they are sent for final trials and registration. Genetic-statistical models that include random and fixed components simultaneously, such as mixed models, have been gaining ground in the current research scenario, as can be seen in the work of (BARETTA et al.,

2016; BOER et al., 2007; SOUZA et al., 2015).

The Blup genotype estimates of the hybrids indicated that KSP22, KSP13 and KSP12 showed high performance for the grain yield character, these genotype averages were higher than those ranked from the 22nd hybrid onwards, among which we can highlight the KSP22 hybrid with a yield of more than 150 sc ha^1 , which is much higher than the national average for grain yield. This hybrid has the potential to be tested in more locations, along with other cultivars that are already commercial, to assess its agronomic performance and if it does prove to be a high performer, it could possibly be launched commercially by the breeding company KSP Ltda.

The maize hybrids KSP1 and KSP3 showed grain yields below those expected for a pre-commercial hybrid. These hybrids will possibly be included in trials in other locations, where they will not necessarily be eliminated from the programme, as perhaps the effects of the Frederico Westphalen environment where they were evaluated penalised their performance. An alternative is to test them in other environments and, if low performance is found, they could be eliminated from the pre-commercial hybrid test group. With regard to the mass of one hundred grains, a trait that shows an association with grain yield (NARDINO et al., 2016; SOUZA et al., 2015), KSP24, KSP7, KSP8 and KSP22 showed the highest magnitudes of the trait. It should be noted that the predicted genotypic values for the trait are in line with other studies, as mentioned above, of the association between grain yield and MCG. The presence of genetic variation between the hybrids studied indicates that there may be gains to be made through genetic improvement of these simple maize hybrids. The pre-commercial maize hybrid KSP22 shows high performance for grain yield, which makes it possible to include it in regional trials with commercial witnesses.

CHAPTER 13

Methods for obtaining hybrid maize seeds

Vinícius Jardel Szareski, Ivan Ricardo Carvalho, Maicon Nardino

Maize *(Zea mays* L.) is one of the most widely grown cereals in the world and has increased grain yields per hectare (KU et al., 2010). According to Menegaldo (2013), functional components present in the composition of maize have begun to be identified in the last five years and are of great importance to the human diet. Brazil is currently one of the largest producers of maize, surpassed only by the United States and China (USDA, 2012). Among the Brazilian states, Paraná stands out as the country's leading producer with 16.75 million tonnes, but the average Brazilian yield is 4,808 kg ha' \ far below the genetic potential of the cultivars and far below the countries that are the largest producers (CONAB, 2013).

Increased research at the beginning of the 20th century made it possible to obtain hybrid maize. Among the various crops improved through genetic improvement, there is possibly no other species of economic importance that has benefited so significantly from scientific research and selection progress (PATERNIANI, 2001). According to Ramalho et al. (2012), genetic improvement through the inclusion of hybrid maize in the production system was responsible for around 58% of this gain. The increase in productivity can also be attributed to improved cultural practices, technification of producers, increased information on how to conduct and manage the crop, among other factors that contributed to boosting the cereal's yield.

Maize *is* currently grown throughout Brazil and *is* the only species among the economically important poaceae that has monoecious floral organisation (SANGOI et al., 2006). According to Fuzatto et al. (2002) and Lorencetti et al. (2005), the success of a maize breeding programme depends on the efficiency with which the genitors are chosen and which, when crossed, produce superior hybrids. However, one of the problems faced by breeders working with line hybrids is assessing the combining ability of the parent lines. Cruz and Regazzi (1994) report that by using the diallel method it is possible to verify the effects of the combining capacities between maize lines. According to Carvalho et al. (2004), the number of genitors used in a diallel is usually quite high, which often makes it unfeasible to cross all the lines, so derivations such as partial diallels arise to test the heterotic effects of two different groups of maize lines.

For Bemini and Patemiani (2012); Cruz and Regazzi (1994); Patemiani and Viegas, (1987) and Vencovsky (1987), the diallel analysis method makes it possible to estimate genetic parameters that are useful for selecting genitors for hybridisation and to check the effects of combining ability, which is characterised as general and specific. General combining ability is characterised by the predominance of additive alleles and the effects of specific combining ability show the effects of non-additive alleles, predominantly in dominance or epistasis. In this context, diallel crosses are important for predicting the best combinations, as they make it possible to direct financial and human resources more efficiently in order to generate constant genetic gain in the development of elite hybrids (VALERIO et al., 2009).

According to Carvalho et al. (2004), in order to maximise efficiency in breeding programmes and reduce the time it takes to obtain elite lines, studies of the association between traits are of great importance as they enable faster progress in plant breeding programmes. In this context, indirect selection becomes an alternative and is recommended for characters that have low heritability, are difficult to measure or when selection is carried out in the first selection cycles. Understanding the associations between traits is extremely important for genetic gains and selecting the best genitors, the ultimate goal of which is to develop superior cultivars with high performance for the main desirable traits (HALLAUER, 2007).

In this sense, the *path analyses* developed by Wright (1921, 1923) are an important tool for breeders, as they break down the correlation or association between the other characters and the character of interest into direct and indirect effects (CRUZ; REGAZZI, 1997; CRUZ; CARNEIRO, 2006), demonstrating the real cause and effect associations between the characters (KUREK et al., 2001), and progress is maximised by using this tool compared to direct selection (FERREIRA et al., 2007).

History and origin of maize

Archaeological studies indicate that maize originated in Mexico (PATERNIANI, 1978). Evidence shows that indigenous populations in Mexico instinctively began domesticating maize based on phenotypes that showed promising characteristics from the population's point of view, where the most vigorous, productive, disease-resistant plants with the best adaptation to climatic conditions were selected. Thus, through a process of continuous selection, a grass with several stalks, small spikelets and few grains evolved into the current maize plant, which was probably domesticated in southern Mexico (MIRANDA FILHO; VIÉGAS, 1987).

Ramalho (2004) reports that although there is some controversy about the evolutionary origin of maize, it is known that its closest wild relative is the teosinto *(Zea mays* ssp. *parviglumis). The* teosinto is a plant with many lateral branches that emerge from its central stem and end in a male inflorescence. Each branch also has a female inflorescence, which gives rise to a small spike. The teosinth has tillering, unlike maize which only expresses this characteristic under biotic or abiotic stress conditions, but with fewer tillers than its ancestor.

More than 300 races and thousands of varieties of maize have been identified and are grown in various parts of the world (TEIXEIRA et al., 2005). The characterisation of the maize kernels of these races in terms of the type of endosperm, 40% are starchy, 30% are hard crystalline kernels, 20% are toothed, less than 10% are popcorn, and around 3% are sweet corn. It is estimated that 100,000 maize accessions are kept in germplasm banks around the world (CHANG, 1992).

The favourable characteristics of the maize cycle and reproductive system have made the crop a model for genetic studies of allogamous species. The characteristics of the monoecious reproductive system and open pollination were mechanisms that favoured the genetic improvement of maize (PATERNIANI et al., 1987).

Botanical description of maize

Maize is a grass belonging to the kingdom Plantae, division Magnoliophyta, class Liliopsida, order Poales, family Poaceae, subfamily Panicoideae, tribe Maydeae, genus *Zea* and species *Zea mays* L., with 2n = 2x = 20 chromosomes (PATERNIANI; CAMPOS, 1999).

According to a review by Buckler and Stevens (2006), five species of the genus *Zea* are currently recognised: Zea *diploperennis - Zea perennis - Zea luxurians - Zea nicaraguensis* and *Zea mays* L.. And within the *mays species* there are four subdivisions which are the subspecies: Zea *mays* L. ssp. *huehuetenangensis', Zea mays* L. ssp. *mexicana', Zea mays* L. ssp. *parviglumis; Zea mays* L. ssp. *mays* (cultivated maize).

Morphological characteristics of maize

The maize crop has an annual cycle and varies in size, with genotypes reaching between 1-4 metres in height. The roots are fasciculated and can be 1.5 to 3 metres long. The leaves of the plants are arranged alternately, supported by the overlapping sheaths that surround the stalk. The plants usually have one to three spikes, with variations depending on fertilisation and management (plant density and spatial arrangement). The female inflorescence corresponds to a stalk with more condensed intemodes, where the flower is partially surrounded by a palea and lemma. A functional pistil is present with a single basal ovary and long styles that are exposed for pollination. The male inflorescence is located in the terminal part of the thatch and has a central axis and several lateral branches. The pollen grains are formed in these structures, which are dispersed and fall into the stigmas, germinating and fertilising the ovule. For most genotypes of maize, the reproductive system is protandric. The pollen grains remain viable for three to five days before the female inflorescence appears on the same plant, and the main mechanism contributing to pollination is the wind, characterising maize as an anemophilous species (PATERNIANI, 1978).

Maize kernels are characterised as the fruit of a seed or caryopsis. The outer layer (pericarp) is derived from the ovary wall and can be colourless, red, brown, orange or variegated. Inside the grain is the endosperm and embryo. The endosperm accounts for approximately 85% of the grain's total mass, the embryo 10% and the pericarp 5%. With the exception of its outermost layer, which consists of a layer of aleurone cells. The embryo is positioned in a depression on the upper surface of the endosperm near the base of the grain. The scutellum consists of a leaf modification that acts as a digestive organ during germination and seedling development. The end of the sprout is surrounded and protected by the coleoptile, while the coleorrhiza surrounds the primary root (PATERNIANI, 1978).

Hybrid maize

The production of hybrid maize began with the work of Beal (1880), who hybridised open-pollinated varieties, pointing to hybridisation as a method of increasing maize productivity. Shull (1909) presented a basic scheme for producing hybrid maize seeds that is still valid today. The scheme is based on obtaining pure lines and using these pure lines to produce hybrid maize seeds. East (1908) carried out parallel work to study the effects of inbreeding and hybridisation in maize. The researcher thought that the pure nail method was not viable from a commercial point of view, due to the low seed production of the strains. Jones (1918) suggested that double hybrids should be used for commercial hybrids, by crossing two single hybrids, making production economically viable. The first commercial hybrids were introduced around 1930 in the region known as *"Corn Belf"* in the United States. Since then, open-pollinated varieties have gradually lost ground. In 1960, the area planted with open-pollinated varieties in the "Corn *Belf*" region corresponded to less than 5% of the total area cultivated with maize (PATERNIANI, 1978).

The introduction of hybrid maize in the 1920s gave a great boost to modern agriculture, and the contributions of maize breeders were very important for the constant increase in productivity. For many years, open-pollinated cultivars were used. East (1936) and Shull (1946) laid the theoretical foundations for obtaining hybrids, but it wasn't until years later that maize hybrids emerged through the use of strains, but due to the low grain yield and the high cost of producing strains, double hybrids dominated the market in the United States from 1934 to 1960. In 1940, 50% of corn crops in the US were hybrids (TROYER, 2006).

With the selection and genetic advancement of the lines, associated with better management conditions and increased productivity of the lines, single hybrids were gaining ground (Figure 1), as the yield results were far superior to the double hybrids that had been developed until then. The gains were even more marked with the development of new technologies, which more recently, with the advent of biotechnology in the mid-1990s, made it possible to increase productivity.

In Brazil, the first maize breeding programme was developed in 1932 at the Campinas Agronomic Institute (IAC) in Campinas - SP. At the IAC, Krug and collaborators produced the first Brazilian double hybrid (KRUG et al., 1943). However, in 1935, Gladstone and Secundino began researching maize at the Federal University of Viçosa, producing the first commercial hybrid in 1938 (PATERNIANI; CAMPOS, 1999).

Hybrid maize production process

Generally speaking, a hybrid production programme consists of the following stages - choosing populations; obtaining strains; evaluating combining ability and extensive testing of the hybrid combinations obtained, traditionally known as cultivation and use value (VCU) testing. Of all these stages, the choice of populations to be self-seeded is of fundamental importance, as the entire success of the programme will depend on it (PATERNIANI; CAMPOS, 1999). It is in these populations that the favourable alleles for the traits of agronomic interest should be concentrated, allowing for the extraction of superior lines (HALLAUER et al., 2010). The choice of populations for the extraction of superior lines is a crucial stage for the success of the breeding programme. There are several options for base populations for the extraction of promising lines for hybrid production. An alternative that is currently being widely used to form base populations is the use of germplasm with a narrow genetic base, such as the use of commercial simple hybrids, which have the advantage of having already been evaluated for desirable phenotypes and tested in various environments, associating high productivity with the large proportion of favourable alleles already fixed in the populations that would be assembled (AMORIM; SOUZA, 2005; OLIBONI et al., 2013).

According to Carvalho (2008), after assembling the population, the most common method used to obtain lines is by selfing the phenotypically selected plants in the population. Self-fertilisation consists of pollinating the ear with pollen from the same plant, leading individuals to homozygosity - in the case of maize, there is a marked loss of vigour in the plants - but in order to improve the lines, some kind of selection process needs to be applied at the same time as increasing the inbreeding of the lines.

Once the collection of elite lines has been obtained, various methods can be applied to assess the combining ability of the pre-selected lines, including complete, partial, circulating and *top cross* dialogues. These methods are based on the concepts of general and specific combining ability. Although the process of obtaining lines may seem complicated and slow, it is not the greatest difficulty in producing hybrids. The process of obtaining superior genetic combinations is extremely difficult, as not all lines will produce superior hybrids due to their low combinatorial capacity with other previously selected lines (CARVALHO et al., 2008).

Below are some of the advantages and disadvantages of using hybrid maize, according to Paterniani (1974):

Advantages of hybrid production:

- Association of characteristics of different parents in the shortest possible time;

- Obtain superior genotypes in a relatively short time;
- Utilising genetic interactions in the hybrid generation;
- Produce uniform genotypes;

-        Achieve less interaction with the environment in the F generation$_x$;

- Producing hybrid maize seeds on a commercial scale, with favourable repercussions for the region's economy.

Disadvantages of hybrid production:

- Only part of the useful genes in the maize will be used, as long as there is no competition for methods to increase the frequency of favourable genes.

- Heterosis is exploited randomly, reaching a ceiling that is difficult to surpass.

- Heterosis can only be used in species where hybrid seed is easily produced.

- Hybrid seed production is only viable where it is easy to process and distribute.

Diallel analysis

The strategies adopted in breeding programmes depend on the genetic analysis of the characters of interest, as they lead to a better understanding of the genetic relationships between the lines involved in the crosses. In this context, diallel crosses are useful for predicting the best combinations between the parents and segregating populations (VALÉRIO et al., 2009; BALDISSERA et al., 2012).

In this way, Sprague and Tatum (1942); Hayman (1954); Griffing (1956) proposed the concept of diallel crossing as the recombination of the genetic variability available within the programme, with the combination of all the genitors, where with n genitors it is possible to obtain n^2 combinations. According to Veiga et al. (2000); Carvalho et al. (2004), the main restriction arising from complete diallel crosses occurs when there are a large number of genitors involved and the number of hybrid combinations to be evaluated is relatively large. These factors cause the inconvenience of high costs, which in many circumstances limit the breeding programme. However, if the researcher's objective is to cross a set of genetic constitutions with one or more testers, partial diallel crossing is used, which is similar to a factorial genetic design that allows crossing between groups rather than within groups. The genetic constitutions must be contrasting and allocated to heterotic groups with characteristics within the unusual group and divergent between the heterotic groups (VENCOVSKY;

BARRIGA, 1992; MIRANDA; GORGULHO, 2001). The partial diallel model was clarified by Kempthome and Cumow (1961), a method developed to increase the number of genitors that can be included in diallel crosses (HALLAUER et al., 2010).

Vencovsky (1987); Bemini and Paterniani (2012) mention that the diallel analysis method makes it possible to estimate genetic parameters that are useful in selecting genitors for hybridisation, such as identifying the gene actions that control the characters, as well as identifying the best combinations of lines to be used as male and female genitors, with the aim of providing maximum heterotic expression in the hybrids. Vencovsky (1987) also mentions that diallel crosses make it possible to obtain estimates of genetic parameters, thus increasing the amount of information for breeders and contributing to decision-making.

Combinatorial capacity

For genetic improvement, it is important to know the genetic relationships found between crosses, which will serve as a decision-making point when choosing combinations. Selection based only on desirable traits is insufficient to produce progenies with high genetic potential. It is necessary to identify genitors with high combinatorial capacity for maximum expression of heterosis in maize hybrids. Diallel analysis is used to estimate the effects of general and specific combining ability (CRUZ; REGAZZI, 1994). The methodology has been widely used in maize breeding and has proved to be efficient in detecting genetic divergences between strains and allocating them into distinct heterotic groups (HAN et al., 1991; GONZALEZ et al., 1997; TERRON et al., 1997), thus demonstrating its great importance in the study and exploration of the best crosses between strains.

According to Sprague and Tatum (1942); Vencovsky (1987); Makanda et al. (2010), general combining ability is associated with additive effect genes and is defined as the average behaviour of a strain in hybrid combinations, and its knowledge is of great importance in the breeding programme for indicating the best crosses and selecting genitors. The specific combining ability is estimated as the deviation of the behaviour of the cross from what would be expected based on the general combining ability of the parents, and is related to non-additive gene effects, which would be in complete dominance, partial dominance and/or epistasis, behaviour that leads certain combinations to be superior or inferior in relation to the average of the lines involved.

Genotype and environment interaction is an extremely important factor and a significant source of variation in maize hybrid trials. There is evidence that the variance of CGC and CEC estimates can interact with locations and years and that CEC includes deviations from dominance and epistasis, as well as a significant portion of the genotype x environment interaction (ROJAS; SPRAGUE, 1952). Sprague and Eberhart (1977) recommend two repetitions per location and three to five environments for evaluating maize crosses, because the interaction of additive effects by environment is a significant factor in the manifestation of phenotypic variance. Increasing the number of environments reduces the error and the interaction of additive effects by environment, while increasing the number of repetitions only reduces the contribution of the error to the phenotypic variance (EBERHART et al., 1995).

Genetic variance

Estimates of the components of variability are extremely important for breeders, as they make it possible to infer how much of the variability is due to genetic and environmental effects (RAMALHO et al., 2012). Allard (1971) described that the variability observed in traits originates from genetic differences and the environment in which the plants grow.

The genetic variance of the character can be obtained from the following formula (RAMALHO et al., 2012):σ

$$\sigma^2_G = \sigma^2_F - \sigma^2_E \text{ ' where;}$$

σ^2_G = variance of genetic origin;

σ^2_F = variance of phenotypic origin;

σ^2_E = variance of environmental origin.

Knapp et al. (1985) used the following models to identify the variation in progenies within each environment:

$$\sigma^2_G = \frac{QMT - QMr}{Rep}$$

σ^2_G : Genetic variance within each environment;

QMT: Mean square of treatment (hybrids) for each environment;

QMr: Mean square of the residue for each environment;

Rep: Number of repetitions of the experiment in each environment.

In this way, all the effects of environment and error are removed, leaving only the genetic effects of the character in the environment studied.

The characterisation of biological variation is the basis for the work of the geneticist or plant breeder, where in order to distinguish and understand the hereditary basis of characters it is necessary to distinguish the two components: genetic and non-genetic (environmental) and the magnitude of their effects. In this context, genetic-statistical models have helped to elucidate the variation present, especially for quantitative traits. The efficiency and clarification of these models has made a significant contribution to the work of plant breeders (BUENO et al., 2006).

Heritability

The study of heritability is important for making practical decisions in crossbreeding methods and is a very useful parameter for breeders. According to Fehr (1987), heritability is conceptualised in a broad and narrow sense. Heritability in the broad sense is provided by the total genetic variance, including additivity, dominance and epistasis in relation to the phenotypic variance. Heritability in the narrow sense is the proportion of additive variance in the phenotype variance. Heritability concepts are important for analysing and identifying traits that fluctuate in different environments, because it allows us to predict the possibility of selection, reflecting the proportion of phenotypic variation that can be inherited. It is obtained from the expression (RAMALHO et al., 2012):

$$h^2 = \frac{\sigma^2_F - \sigma^2_E}{\sigma^2_F} \times 100$$

where,

h^2 = heritability of the character;

σ^2_F = is the phenotypic variation;

σ^2_E =is the variation in environment;

According to Ramalho et al. (2012) estimates of the heritability of a trait are not immutable, because the trait varies in relation to the genetic variation present and the effect of the environment. In this context, heritability estimates can be increased by using populations with greater genetic variation and controlling the effects of the environment through better experimental conditions, as a way of reducing the environmental contribution to total phenotypic variation.

Due to the complex nature of the gene-character relationship, it is expected that environmental factors can obscure certain inheritance models, even if these factors are subtle and trivial. Most genes are expressed in a fairly uniform and predictable way, considering the normal environmental conditions to which they will be exposed, the expressivity of a gene can be markedly altered by genetic and/or environmental factors. The expressivity of a gene refers to its mode of expression, which generally fluctuates under different environmental conditions. Penetrance refers to the expression or non-expression of a gene. In this way, penetrance and expressivity are phenomena that mask the gene-character relationship and tend to obscure the correspondence between genotype and phenotype, making the task of the plant breeder more difficult, especially for characters with incomplete penetrance and variable expressivity (ALLARD, 1971).

Correlations between characters

Correlation is a statistical parameter that measures the degree of association between two variables and/or traits. Variables are considered correlated when variation in one is accompanied by simultaneous variation in the other (RAMALHO et al., 2012). The effects of genetic correlation are explained by the genetic mechanisms of the joint variation of two variables, with the effects of pleiotropy and genetic linkage. Knowledge of the phenotypic, genetic and environmental correlation between traits can be essential for the simultaneous selection of several traits, or when the trait of interest has low heritability or is difficult to measure (FALCONER, 1996). According to Carvalho et al. (2004), in order to maximise the efficiency of breeding programmes, great importance should be attached to studies of correlated characters, as they enable breeders to make faster progress than by selecting the desired character directly.

According to Cruz and Regazzi (1994), phenotypic correlations can be directly measured by measuring two characters in a certain number of individuals in the population. This correlation has genetic and environmental causes, but only the genetic ones involve an association of an inheritable nature, which can be used to guide plant breeding programmes. Therefore, in genetic studies it is essential to distinguish and quantify the degree of genetic and environmental association between characters. According to Ramalho et al. (2012), genetic correlations occur because when two genes are weakly linked they tend to be transmitted in the same gamete, giving rise to an individual that may or may not express the phenotypes corresponding to the inherited alleles; a similar situation occurs when the gene in question is pleiotropic, in which case the phenotypes associated with the same allele will always tend to occur together.

In order to better understand the association between characters, Wright (1921) proposed the method known as *path analysis,* which breaks down the estimated correlations into direct and indirect effects of each character on a basic variable, and makes it possible to assess whether the correlation between two variables is cause and effect or generated by the influence of other variables.

According to Souza et al. (2008), the correlations between the primary components and the secondary components of production can provide different selection strategies, leading to growing increases in genetic progress with selection cycles for agronomic traits of interest. Magalhães et al. (2002) pointed out that maize grain yield and primary production components correlate positively with germination uniformity and speed, prolificacy, total number of leaves, number of leaves above the ear and production efficiency, and negatively with number of days to flowering, lodging, number of branches on the stalk, ear height/plant height ratio and leaf insertion angle.

Maize is one of the most extensively researched crops in the field of genetics, and serves as a template for many other allogamous species. It is important not only because of this, but also because of the large quantity of food available to the entire production chain, as well as its recent inclusion in the functional food market.

It is because of the importance of maize not only at national level, but also worldwide that research continues to be carried out efficiently and reliably by the countless researchers who work in this area. Genetic improvement, together with other areas, has progressively contributed to the genetic advancement of maize, but the former, year after year with technological innovations and the insertion and adherence of molecular practices, is mainly responsible for the genetic gain and the genetic levels of hybrids achieved today, Of course, the grain yield is far below what the hybrid's genetics can really generate, but little by little cultural practices and technological levels are being improved in the most varied production systems, reducing costs and thus facilitating the adherence of small producers, who contribute significantly to production at a national level. A great deal of research has been and is being carried out, and more and more traits that until recently were considered to have little influence are now raising doubts among breeders, especially about the extent to which they are related to grain yield in maize.

CHAPTER 14

Seed production of open-pollinated varieties

Ivan Ricardo Carvalho, Vinícius Jardel Szareski, Maicon Nardino

Population genetics provides the basis for breeding. All populations have a gene pool that is particular to them and that characterises them, which is passed on generation after generation. Within a given population, breeding works to identify the individuals who possess characteristics that are desirable from an agronomic point of view, and indicates the frequencies of these alleles, which will be passed on to subsequent generations.

The probability of obtaining an agronomically desirable strain will depend on the frequency of favourable alleles in the population (Duvick, 2005; Troyer & Wellin, 2009). These frequencies can be increased through different selection methods, such as mass selection and/or intra- or inter-population recurrent selection (Vencovsky, 1987). One of the main ways of knowing the additive and non-additive genetic components in the genetic control of traits is through the diallel crossing method, where they occupy a prominent place in the genetic improvement of maize (Jesus et al., 2008; Souza et al., 2008; Bueno et al., 2009; Faria et al., 2015; Paini et al., 2015).

The length of a crop's cycle is a character that is much worked on by breeders and requires a great deal of attention during the selection phases in populations. In order to understand the genetic control of flowering in maize, Koester et al. (1993) identified that the number of days to reach flowering is controlled by quantitative loci located on chromosome 1, close to the centromere region. The maize crop has three separations in terms of cycle length, which can basically be classified as super early, early and late (Sangoi et al., 2002). Fancelli & Dourado Neto (2004) classify super-early cycle constitutions as those requiring between 780 - 830 calorific units, early ones between 831 - 890 calorific units and individuals classified as late cycle have a calorific demand of between 891 - 1200 calorific units until they reach flowering.

Most traits of economic interest show continuous segregation, such as grain yield, plant height and length of the vegetative cycle, where in a population they can show wide variation (Hallauer et al., 2010).

The multiple factor hypothesis explains that the control of a quantitative trait is influenced by a large number of genes, each of which contributes a small effect to the phenotype, so the variations found for quantitative traits are much greater, which suffer a marked effect from the environment, a factor that demands attention from geneticists and breeders, as it is necessary to break down the total variation into genetic and environmental effects (Ramalho et al., 2012).

Studies by Lima et al. (2008) reported the genetic control of flowering in homozygous maize lines, but there is little information in the literature about the genetic actions that control the cycle in creole maize populations. Within this context, the aim of this study was to evaluate segregation and reciprocal effects, relating them to the predominant genetic effects on the requirement for thermal sum in crosses of populations contrasting in terms of thermal requirement.

The creole maize populations were collected in the municipality of Seberi - RS (27°28'20"S and 53°24'10"W) in February 2010, namely: population 1, population 2 and population 3. The experimental work was carried out in the 2011/2012 and 2012/2013 agricultural harvests. In the 2011/2012 harvest, the populations were sown to obtain seeds for the hybrid populations. In the 2011/2012 harvest, the three parent populations were sown together with the hybrid seeds from the crosses.

The region's climate is classified as subtropical Cfa and has a sub-humid climate, with an average annual temperature of 18.8°C, the average temperature of the coldest month of 13.3°C and the temperature of the hottest period ranging from 18 to 26°C. The soil was prepared conventionally with two harrowings, and the sowings were made on red alumino ferric latosol (Streck et al., 2008).

The design used was a randomised complete block design consisting of six eight-metre rows spaced 0.3 metres apart and 1.0 metre apart. The four centre rows were considered for evaluation. In both experiments, the fertiliser used at sowing was equivalent to 300 kg ha^1 of 05-20-20 chemical fertiliser and 135 kg ha^1 of nitrogen in the form of urea at stage V6, in the two years of the experiment.

To determine the cycles of the populations, the average thermal sum for 50% of the individuals in each population and in the subsequent hybrid populations was taken into account, i.e. when 50% of the individuals showed male flowering, considering the period between sowing and flowering.

The cycle of the populations and their subsequent hybrid populations were classified according to the methodology presented by Fancelli & Dourado Neto (2004). According to this methodology, the cycles of the populations were identified: population 1 with 824.86 GD is classified as super-early, population 2 with 874.89 GD is classified as early and population 3 with 1032.39 GD is classified as late. Based on the characterisation of the populations with contrasting cycles, the crossing blocks were set up with the staggered sowing of individuals from each population, in order to cover as many crosses as possible.

The crosses between the populations were made in a complete diallel scheme, forming six combinations of the populations, 1x2, 1x3, 2x1, 2x3, 3x1 and 3x2, each cross consisting of five female x five male plants. The crosses were carried out artificially by protecting the female inflorescence before the stigmas of the female plants emerged. Pollen was collected from the male plants by protecting the male inflorescence with a paper bag approximately 24 hours before fertilisation. After the artificial crosses were made and the seeds collected in the 2010/2011 harvest, the original populations and the hybrid populations were sown in the following harvest, in 2011/2012. Evaluations and data collection were based on the period between emergence and flowering.

The base temperature for maize was 10°C for the entire crop cycle, as recommended by Berlato & Matzenauer (1986). Equation 1 (Amold, 1960) was used to determine the daily thermal sum.

$$GD = \sum_{i=0}^{n} \to (\frac{T\,max + T\,min}{2} - Tb)\ (1)$$

GD is the total accumulated degree days from emergence to flowering; Tmax is the maximum daily air temperature (°C); Tmin is the minimum daily air temperature (°C); Tb is the base temperature (°C); n is the number of days from sowing to female flowering. The accumulated thermal sum (STa °C day) from sowing to female flowering was calculated as STa=SGD.

In the frequency distribution analysis for the diallel, all the crosses had different mean, median and mode values,

Asymmetric distributions were observed for the thermal sum. With regard to the magnitude of the asymmetry, the crosses 2- 3x1; 3- 1x2; and 6- 2x3 had positive asymmetry, and the individuals were located to the right of the distribution curve. For crosses 1- 2x1, 4- 3x2 and 5- 1x3 the asymmetry was negative, so the individuals were concentrated to the left of the distribution curve. Averages very far from the mode contribute to a greater extension of the tail, with the averages of crosses 1- 2x1 and 2- 3x1 being furthest apart, so the tail of the distribution assumes a greater extension in the distribution and a greater contribution to the asymmetry of the individuals in these crosses. Crosses 1, 2 and 3 revealed greater variance within the cross, while crosses 4, 5 and 6 revealed lower magnitudes of variation, meaning that the individuals showed thermal sum values with less variance and were closer together.

The 2x1 cross (super early father x early mother) revealed that the resulting populations showed a tendency to increase the thermal demand compared to the parents, with 80.58% of the plants having a higher thermal demand compared to the parents, where the highest frequency of individuals is for the thermal sum of 912 - 924 °C with 23.30% of the individuals. The populations resulting from crosses between super-early (father) and early (mother) cycle populations increase the vegetative cycle due to the higher thermal demand, indicating that the gene actions for the vast majority of the group are overdominant. According to Hallauer et al. (2010) the increase in the thermal sum in progenies points to the existence of heterosis.

With regard to the 2-3x1 cross between the super-early (male donor) and late (female recipient) cycle populations, the hybrid population had a heat sum between the values of the two parents. The highest frequency of individuals was between 915-930°C, with an overall average of 938.35°C. In this cross there is a strong indication that the genetic control was only the additive component. Câmara et al. (2007) worked with two populations of tropical maize and evaluated the genetic parameters involved in characters of agronomic interest and found high heritability estimates (94.12%) for male flowering. These estimates facilitate the genetic improvement of hybrid populations for male flowering.

In the 3- 1x2 cross between the super-early (female recipient) and early (male donor) cycles, the hybrid population tended to have a higher heat sum requirement. Only 5.15 per cent of the individuals were located between the average of their parents, while 13.40 per cent of the individuals had the same heat sum requirement as their father. In this case, the effects of overdominance were more pronounced, resulting in a population with most of the individuals superior to the largest (early) parent, i.e. overdominance resulted in a late hybrid population. The use of earlier varieties allows the area to be cleared earlier, increasing exploitation, but this is not always synonymous with higher yields, since early varieties have less time to recover from environmental limitations, such as water stress (DE ARAÚJO et al., 2013).

In the 4- 3x2 cross between late-cycle (female recipient) and early-cycle (male donor) parents, the hybrid showed lower thermal requirements than the parents, with 22.85 per cent of the individuals being between the average of the parents. For genetic improvement, these results are of great relevance, as they indicated that crossing an early variety with a requirement of 874.9±2.4 with a late-cycle variety of 1032.4±2.4 resulted in a hybrid population with a lower requirement for thermal sum than the populations used as parents. According to Ramalho et al. (2012), it is important to emphasise that interactions of dominance and/or overdominance do not always act to increase the phenotypic value, interactions can occur which reduce the expression of the character, this was evidenced by the results obtained in this cross by Araújo et al. (2013) evaluated the agronomic performance of creole varieties and maize hybrids grown in different management systems, where they observed that there is a decline in creole varieties when the technological level is increased, however the varieties favour autonomy in the production process and according to their results they observed that the creole varieties were able to respond to the technological increase in farming. In this way, the development of super-early or early hybrid populations, as obtained in this work, associated with satisfactory yields, should be pursued because many farming systems still depend on these creole varieties.

With regard to the 5- 1x3 cross, with the female population characterised as super early (824.86+3.9) and the male population characterised as late cycle (1032.4+3.9), the resulting hybrid population showed segregation between the two parents (figure 5). According to Ramalho et al. (2012) these effects are explained by additive genetic interaction, in which each allele contributes a small cumulative effect on the phenotype, where the average of the Fi generation is equal to the average of the parents.

With regard to the 6- 2x3 cross, where the early-cycle population (874.89+1.6) and a late-cycle parent (1032.39±1.6) are used as the mother, the hybrid population showed segregation between the average of the parents, with 15.27 per cent of the individuals showing an average thermal demand equal to the male parent requiring the greatest thermal sum, revealing that in this case the effects of additivity are more pronounced. This was different when the early male and late female populations were crossed, where the hybrid population showed a tendency towards lower heat requirements. When crossing an early mother and a late father, the hybrid population resulted in values between the two genitors, but the results of this cross are similar to crosses 2 and 5. Improving creole populations, especially for a shorter cycle, is important for targeting regions with low cultivation technology, as they already have several advantages linked to resistance to diseases, pests and climate fluctuations, and because their seeds can be stored for subsequent harvests, reducing the cost of production for farmers (Carpentieri-Pípolo et al., 2010; Ferreira et al., 2009; de Araújo et al., 2013).

The results of the crosses between contrasting populations for the vegetative cycle character revealed that the effects on the hybrid populations are different when the original populations are used as the father or mother in the cross, so it was deduced that there was a maternal effect. In this sense, the reciprocal crosses made it possible to elucidate the effects of the population when used as a male or female, indicating that the inheritance of the character is controlled by nuclear genes.

Most traits in higher plants are controlled by genes in the nucleus, but there is another group of traits that are inherited from genes in the cytoplasm. Therefore, if the character is controlled by cytoplasmic genes, the mother is responsible for its expression in the offspring. If the results of the cross and its reciprocal are different, the expression of the character may be due to the effects of cytoplasmic genes, which persist in successive generations and/or the phenotypic effect of maternal origin and the product of the cross will only manifest itself in the following generation (Ramalho et al., 2012).

According to the results of this study, both additive and non-additive gene actions are important in explaining the variation in hybrid populations in terms of heat sum, results that corroborate Ajala (1992) where both additive and non-additive gene actions were important in explaining flowering in maize. In studies with four strains in a diallel crossing scheme, Lopes et al. (1995) concluded that the cycle character is controlled by a few genes and its control occurs through additive and dominant genetic actions, data that also corroborates this work, in the study of the segregation of the cycle of Creole populations in diallel crossing. Lima et al. (2008) used two lines and evaluated Fi, F_2 , RCi and RC_2 , concluding that the genetic control of flowering in maize is due to additive and dominant components.

Reciprocal effects must be taken into account when selecting populations with regard to heat requirement. Additive and non-additive gene actions are important for the heat sum character when contrasting populations are crossed for heat requirement.

In order to reduce the cycle, it is necessary for the population used as a female to have a late cycle and the one used as a male to have an early cycle. Crosses between precocious female and super precocious male or super precocious female and precocious male populations generate populations with higher thermal requirements.

CHAPTER 15

Genetic improvement and management of seed production fields

Maicon Nardino, Ivan Ricardo Carvalho, Vinícius Jardel Szareski,

Plant breeding is a combination of art and science. It is a process that seeks to genetically alter plants in order to meet the world's growing demand for food, in a way that serves humans and animals.

The need for food in the future will be much greater than it is today, due to the growing population, with the projection that by 2050 there will be more than 9.4 billion people in the world. The risk of world hunger is not new. Around 200 years ago, an Englishman called Thomas Robert Malthus made some bold predictions, postulating in 1798 that human populations would be limited by hunger. Malthus believed that the population would grow geometrically, while the proportion of food would grow arithmetically.

The catastrophic Malthusian prediction didn't actually happen, even though the population increase materialised as predicted. The prediction was only inaccurate thanks to the increase in food production, which was underestimated at the time. The increase in production was due to the advance of new agricultural frontiers, the inclusion of new techniques and inputs. The mechanisation of agriculture with the use of inputs and new cultivars developed through genetic improvement resulted in productivity gains that Malthus did not foresee.

Currently, several countries have large-scale food production, but many regions are subjugated to hunger and human malnutrition due to the precarious distribution of food. The aim of this literature review was to look at the history of plant breeding, focusing on maize, highlight the main breeding methods used and report on the main future prospects.

General objectives of plant breeding

Plant breeding is related to and seeks out various agronomically and economically important traits. The aim is to identify and select superior genotypes, with the aim of obtaining a plant ideotype. However, in the course of this search, some difficulties are encountered, which will be supplemented in part by this work.

The ever-increasing genetic progress of cultivars makes breeding the main pillar for sustaining agriculture. Making fruit, vegetables, cereals and oilseeds available to the human population. It works to increase the production of oils, with more productive species and to increase the production of ethanol and biodiesel. For animals, it improves the quality of forage species, producing more meat, milk and dairy products.

Plant breeding can make production systems more sustainable. With the introduction of genetic materials that are resistant and tolerant to biotic and abiotic factors, there is no need to use insecticides, herbicides and fungicides, which are major precursors to increasing production costs. Even for abiotic factors that are often beyond the grower's control, cultivars have been developed that tolerate fluctuations in temperature and rainfall.

Plant breeding works to create new products with different added values, bringing information from research laboratories to farmers and consumers, among other members of production systems, through the commercialisation of improved seeds and seedlings.

In general, plant breeding is involved in a large context that aims to increase the quality and quantity of specific products, such as oils, proteins and minerals; cultivars that are resistant and tolerant to soil and climate conditions; better preservation of post-harvest products; creating food security for the population and greater opportunities for rural producers to add income.

Areas of knowledge contextualised with plant breeding

The plant breeder must have great skill in selecting plants within a population, depending on the apparent characteristics and the objectives set out in his mind as to which characteristics an ideal plant must possess in order to be selected or remain part of the programme.

In the past, plant breeding followed criteria without scientific methodologies and was carried out as an art form. With scientific advances and the formation of a more consistent base for plant breeders, the selection process has become more rigorous and effective, becoming more of a science than an art. The interrelationship of breeding, especially in current times, with other related areas has brought great scientific advances, especially at the genetic level.

Among the areas most closely related to plant breeding are;

Genetics: knowledge of the principles of gene and chromosome expression, expression of the degree of kinship and choice of the best breeding method to obtain a cultivar.

Biochemistry: study of the actions of proteins and enzymes, as well as their relationship with other molecules, resulting in the selection of individuals with better quality, especially nutritionally.

Physiology: knowledge of the processes of plant growth and development, the relationship with hormone production under specific conditions, helping to identify the response of genotypes to specific environmental conditions.

Botany: an understanding of botany allows us to understand the reproductive systems of plants, identifying those individuals that are superior at leaving offspring for future generations.

Statistics and Experimentation: this knowledge enables the evaluation and identification of genetically superior groups within a group of genetic materials, as well as analyses of kinship, adaptation to different environmental conditions and combining abilities, based on the probability of identifying that certain individuals are superior to others.

Entomology: knowledge of the relationships and anatomy of insects, as well as how insect pests attack plants and identifying the response to the form of attack.

Phytopathology: understanding how the fungus is infected and its aetiology, in order to identify possible resistance in plants.

The related areas of knowledge constitute the tools with which the plant breeder must be equipped to carry out his activities. A few areas have been mentioned that the plant breeder should be aware of, but depending on the objective, other sciences become fundamental, such as soils, plant nutrition, cytogenetics, climatology and agronomy in general, as well as the demands of producers and the market for developing cultivars.

The genetic improvement of plants is involved in a large context of other sciences, so it is a process that must remain continuous, because natural circumstances and needs have been changing over time, so the lines currently much discussed in numerous events, mainly sustainability, must progress and evolve the improvement programmes, so that the preferences of the market are always met, as well as the people who are involved in the different production systems.

Losses caused by genotype x environment interaction

Over the course of improving cultivated species, agriculture has moved towards greater control of environmental conditions. The environment is made up of all those factors that positively or negatively affect the growth and development of plants, i.e. all the components except the genetics of the material.

Brazil currently has a very wide range of cultivars and varieties available on the seed commercialisation market. This large number of genotypes is necessary because of the different responses to environmental conditions. Some materials perform better in conditions of higher humidity and less weeds. While others are more responsive to conditions of higher solar radiation, low humidity and are more tolerant of stress from biotic factors. However, due to these and countless other factors, genotypes vary, especially in terms of productivity due to environmental conditions.

A cultivar may have high technology and genetic potential, but if the environmental conditions are not favourable for expressing its potential, it may not express significant response values to the conditions it is subjected to. Therefore, genetics and the environment must go hand in hand in order to find the total magnitude of the genotype's response, seeking to maximise the crop's genetic potential.

The G x E interaction is an extremely important phenomenon that challenges the role of the plant breeder and the work of the agronomist, especially in cultivar competition tests and recommendations. When genotypes are evaluated in different years and locations, significant results are usually found for the G x E interaction, as the mainly abiotic conditions are contrasting, which changes the response of the same materials under these conditions.

According to Bradshaw (1965), the magnitude of differentiated responses of genotypes to environmental conditions is due to phenotypic plasticity, where depending on the aggregate genetics of the material there may not be significant responses to contrasting environments. These cultivars can undergo a process of phenotypic adjustment that is inversely proportional to the magnitude of heterozygosity. When a cultivar is made up of a large number of genotypes, especially those that have already adapted to the environment, it shows the plasticity of phenotypic adjustment to environmental conditions, due to the "buffering" effect.

Main factors causing significant effects for genotype x environment interaction (g x e)

The main influencing factors are as follows:

> Rainfall during the crop cycle;

> Photoperiod;

> Solar radiation, especially for materials responsive to thermal sum;

> Sowing time - always respecting agro-climatic zoning;

> Insect pest control;

> Disease control;

> Weed control;

> Air and soil temperature;

> Agricultural crop management practices;

>	Soil fertility;

In the search for sustainability in a programme or by the grower himself, these practices are essential to observe, but some of them are difficult to control and are unpredictable, such as rainfall, relative humidity in the air and soil, photoperiod and solar radiation, among others, However, the other practices can be controlled or predicted, and it is important that they take place within the experimental planning or at crop level so that the cultivar reaches or is close to its full genetic potential for productivity and thus has a return on the sale of seeds to the market and also a good return on the producer's earnings.

Main characteristics of maize *(Zea mays* 1.) genetic improvement

There are currently numerous genetically improved species in Brazil, the main ones being maize, soya, wheat, oats and cotton, among others. Among the main commercial crops, maize, with its hybrid technology on the market, is at a very advanced level. Knowing its importance to the market and the entire national scenario, the main factors surrounding the improvement of maize hybrids will be elucidated.

The search for highly productive hybrids aims to take advantage of the effects of heterosis. Heterosis is the genetic expression resulting from the beneficial effects of hybridisation. Inbreeding, another extremely important process in the search for vigour, although it removes vigour from plants, consists of crossing individuals with a degree of kinship, leading to homozygosity of the genetic materials.

Improving strains

Self-fertilisations are essentially important for obtaining inbred lines. The plants in an open population of maize are heterogeneous, essentially simple hybrids, as they result from the union of two different gametes. To this end, there are a number of techniques that are used effectively to control pollination (Figures 1 and 2). The methods consist of protecting the male and female organs (pendant and doll).

There are various methods for obtaining strains, some of which will be described below;

Standard Method - A technique that consists of selfing, with selection between and within progenies as the generations progress. Plants are selected on the basis of favourable agronomic characteristics.

Single pit method - This method is similar to the standard one, but differs in the establishment of the plants, as each progeny is represented by a single pit with three plants. The advantage lies in the reduction in the area occupied by the progenies.

Genealogical method - Two strains with good combining ability are chosen first, with the aim of isolating new second cycle strains from the hybrid between the first strains resulting from inbreeding.

Critical Hybrid Method - This is a recent method which consists of an early combination test through crosses between individual plants. Some authors claim that this method is efficient for developing lines and hybrids.

Zygotic selection - This is also the most up-to-date method. It is adopted when a superior commercial strain is available and the goal is to obtain a new strain from a heterogeneous population, with the aim of obtaining a superior hybrid.

There are other methods that are applied to maize improvement, in general these are some of the two most commonly used, however the best method varies depending on the objective sought by the plant breeder.

Types of hybrids

The types of hybrids present are classified in Table 1.

TABLE 1 - Types of maize hybrids and their respective crosses (Viégas and Miranda Filho, 1978).

Hybrids	Obtaining by crossbreeding
"Top Cross	Lineage x Cultivar
Single Hybrid	Lineage A x Lineage B
Double Hybrid	(Lineage A x Lineage B) x (Lineage C x Lineage D)
Triple Hybrid	(Lineage A x B) x Lineage C
Modified Simple Hybrid	(Lineage A x Lineage A') x B
Multiple Hybrid	Strains (AxB) x (CxD) x (ExF) x (GxH)
Intervarietal Hybrid	Variety A x Variety B

The commercial importance of each hybrid is described below.

1.   "Top Cross" - Commercially it has no value, it is more used in the

breeding programme to evaluate strains that could potentially produce hybrids.

2.   Single Hybrid - This has the greatest potential for productivity, with uniform agronomic characteristics. However, seed production costs are high.

3.   Double Hybrid - Until recently this method was widely used, but with the advent of single hybrids it has lost a little ground in the market. It is generally more stable than single hybrids.

4.   Triple Hybrid - This is the result of crossing a single hybrid with a strain, and its characteristics include good plant uniformity. It takes two years of cultivation to obtain.

5.   Modified Single Hybrid - With a lower seed production cost, this hybrid has taken the place of the double hybrid, as it has a higher degree of heterosis and is less labour-intensive to obtain.

6.   Multiple Hybrid - Not very popular commercially. By using six to eight strains, it has the advantage of greater adaptation.

7.   Intervarietal hybrid - Obtained by crossing different classes of hybrids. Nowadays, with hybrids obtained from self-bred lines, these materials have lost their place, but they were important in the first quarter of the 20th century.

Advantages and disadvantages of growing hybrids

The production of hybrids is not only aimed at beneficial effects, there are also some factors that are important to evaluate.

Main advantages of hybrid seed production:

- Production of highly productive genetic materials;

-Uniformity of   cycle and agronomic characteristics of the plants;

-Use of   gene interactions in hybrid production;

-Materials   responsive to environmental conditions;

Main disadvantages of hybrid seed production:

-The   production of these seeds *is*   only viable in places that have a good structure for processing and distribution.

-The need for isolated fields to produce hybrids in order to avoid contamination.

-Some hybrids have high seed production costs.

-Response fluctuations occur depending on the intrinsic conditions of the environment.

As already mentioned in this part of the chapter, hybrid production involves a series of factors that are extremely important to evaluate. Maize breeding is very advanced in technological terms, but yields are far below the genetic potential of the cultivars. Among the determining factors are sowing times outside the agro-climatic zoning, the recommendation of cultivars that are not specific to the technological conditions of the producer, as well as the use of materials that are not specific to the environmental conditions.

Breeders work to improve cultivars, making them increasingly productive and responsive to soil and climate conditions. However, within the context of sustainability, the whole system will only be efficient and flow normally if all the practices are appropriate and act together, so that both the breeder and the producer will have satisfactory returns within the production system. When looking for sustainability in a system, one must look at the whole environment, not just at specific parts, or those that are said to be more important, because sustainability is involved in production environments by a very large context and only with joint actions and in a reliable way will it be achieved.

Prospects and the future of plant breeding

For many crops such as maize, classical genetic improvement has achieved very high levels of productivity and improved other characters that are correlated and agronomically important.

Developing new cultivars is a time-consuming process that requires the breeder to be careful and have knowledge of many related areas. However, the work that is currently being carried out, especially with the study of strains and hybrids, requires less time. These factors are appearing thanks to advances in genetics and breeding, but also with the addition of other areas and the joint study of these related areas, great progress has been made.

Evolution is taking place today with studies at the DNA level. The combination of these areas has created a new study tool, biotechnology. This tool has collaborated with advances in plant breeding, as well as in other areas. Together with the technique of molecular markers, it has been possible to progressively increase and improve agronomic characteristics by studying plant adaptation responses to environmental conditions at the DNA level. The way in which genes are expressed according to the conditions of the environment, when they are expressed, where and the magnitude of the variation they undergo when the plants are subjected to adverse conditions have been evaluated.

BIBLIOGRAPHICAL REFERENCES

AACC American Association of Cereal Chemists. **Approved methods of the American Association of Cereal Chemists.** lO.ed. Saint Paul: Approved Methods Committee, 2000.

AACC American Association of Cereal Chemists. **Approved methods of the American Association of Cereal Chemists.** 9.ed. Saint Paul: Approved Methods Committee, 1995. v.1-2.

AHLOOWALIA, B. S; NICHTERLEIN, K. Global impact of mutation- derived varieties. **Euphytica,** v.135, n.2, p. 187-204, 2004.

AITA, C & GIACOMINI, S.J. Decomposition and nitrogen release of crop residues from single and intercropped ground cover plants. **Revista Brasileira de Ciência do Solo,** v.27 n.4 Viçosa, 2003.

AITA, C.; BASSO, C.J.; CERETTA, C.A.; GONÇALVES, C.N. DA ROS, **C.O.** Soil cover plants as a source of **N for** maize. **Revista Brasileira de Ciência do Solo,** v. 25, p. 157-165,2001.

AJALA SO. Combining ability for maturity and agronomic traits in some tropical maize (Zea mays L.) populations. **Tropical Agriculture,** v. 69, n. 1,p. 29-33,1992.

ALCANTARA NETO, F.; GRAVINA, G. A.; MONTEIRO, M. M. S.; MORAIS, F. B.; PETTER, F. A.; ALBUQUERQUE, J. A. Trail analysis of soya bean yield in the Alto Médio Gurguéia micro-region. **Comunicata Scientia,** v. 2, p. 107-112, 2011.

ALLARD, R. W. **Principles of plant breeding.** São Paulo: Edgard Blucher, 1971. 38 Ip.

ALMEIDA, A.M.R.; FERREIRA, L.P.; YORINORI, J.T.; SILVA, J.F.V.; HENNING, A. 1995. **Soya bean diseases** (Glycine max). In: KIMATI, H.; AMORIM, L.; REZENDE, J.A.M.; BERGAMIN FILHO, A.; CAMARGO, E.A. **Manual de fitopatologia: doenças das plantas cultivadas.** São Paulo: Ceres, Chap.64, p.569-596.

ALMEIDA, R. D. de; PELUZIO, J. M.; AFFÉRRI, F. F. Phenotypic, genotypic and environmental correlations in soya grown under irrigated floodplain conditions, southern Tocantins. **Bioscience Journal,** Uberlândia, v. 26, p. 95-99, 2010.

ALMEIDA, R. D. de; PELUZIO, J. M.; AFFÉRRI, F. S. Genetic divergence between soya cultivars under irrigated floodplain conditions in southern Tocantins State. **Revista Ciência Agronómica,** v. 42, n 1, p. 108-115, 2011.

ALVA, A. K.; PARAMASIVAM, S.; FARES, A.; DELGADO, J. A.; MATTOS JÚNIOR, D.; SAJWAN, K. Nitrogen and irrigation management practices to improve nitrogen uptake efficiency and minimise leaching losses. **Journal of Crop Improvement, Binghamton,** v. 15, n. 2, p. 369- 420, 2006.

ALVAREZ, C.G.D.; PINHO, R.G.V.; BORGES, I.D. Evaluation of bromatological characteristics of maize forage at different sowing densities and row spacings. **Ciência e Agrotecnologia,** v. 30, n. 3, 2006.

AMADO, T.J.C.; MIELNICZUKJ.; AITA, C. Nitrogen fertiliser recommendation for maize in RS and SC adapted to the use of soil cover crops, under a no-till system. **Revista Brasileira de Ciência do Solo,** 26:241-248, 2002.

AMORIM, E. P.; SOUZA, J. C. Inter- and intra-population maize hybrids obtained from S0 populations of commercial simple hybrids. **Bragantia,** Campinas, v.64, n. 3, p.561-567, 2005.

ANDERSSON, A.; JOHANSSON, E.; OSCARSON, P. 2004. Post- anthesis nitrogen accumulation and distribution among grains in spring wheat spikes. **The Journal of Agricultural Science,** v. 142, n. 05, p. 525- 533,

ARAÚJO, J.L.; FAQUIN, V.; VIEIRA, N.M.B.; OLIVEIRA de, M.V.C.; SOARES, A.A.; RODRIQGUES, C.R.; MESQUITA, A.C. Rice growth and production under different nitrate and ammonium ratios. Red de Revistas Científicas de América Latina, el Caribe, Espana y Portugal. **Revista Brasileira de Ciência do Solo,** v. 36, n 3, p. 921-930, 2012.

ARF, O. FERNANDES, R.N.; BUZETTI, S. RODIRGUES, R.A.F.; SÁ, M.E.; ANDRADE, J.A.C. Soil management and timing of nitrogen application on maize development and yield. **Acta Scientiarum Agronomy.**Maringá, v. 29, n. 2, p. 211-217, 2007.

ARNHOLD, E. et al. Prediction of genotypic values of maize for the agricultural frontier region in northeastem Maranhão, Brazil. **Crop Breeding and Applied Biotechnology,** v. 12, n. 2, p. 151-155, jun. 2012.

ARNOLD CY. Maximum-minimum temperatures as a basis for computing heat units. **Proceedings of the American Society for Horticultural Sciences,** v. 76, n. 9, 682p, 1960.

ASHRAF, M. T.; SHARIF A.; JAMIL, T. Analysis of Variance and Influence of Number of Grains Per spike on Protein Percentage and Yield in Wheat Under Different Environmental Conditions.Pakistan. **Journal of Biological Sciences,** v. 6, n. 4, p.382-385, 2003.

AURORA, Y.; MULONGOY, K.; JUO, A.S.R. Nitrificationand mineralisation potentials in a limed Ultisolinthe humid tropics. **Plant and Soil,** Dordrecht, v.92, p.153-157, 1986.

BAHRY, A.C.; VENSKE, E.; NARDINO, M.; FIN, S.S.; ZIMMER, P.D.; SOUZA, V.Q. de; CARON, B.O. 2013. Urea application in the reproductive phase of soya and its effect on agronomic characters. **Agricultural Technology & Science,** v.7, p.9-14.

BALBINOT, A.A.; BACKES, R.L.; ALVES, A.C.; OGLIARI, J.B.; FONSECA, J.A. Contribution of yield components to grain yield in open-pollinated maize varieties. **Revista Brasileira Agrociência. V.ll,** n. 2, p. 161-166, Apr-Jun, 2005.

BALDISSERA, J. N. C.; VALENTINI, G.; COAN, M. M. D.; ALMEIDA, C. B.; GUIDOLIN, A. F.; COIMBRA, J. F. M. Combinatorial capacity and reciprocal effect in bean agronomic traits. **Semina: Ciências Agrárias,** Londrina, v. 33, n. 2, p. 471-480, 2012.

BARBER, S. A.; MACKEY, A. D.; KUCHENBUCH, R. O.; BARRACLOUGH, S. Effect of soil temperature and water on maize root growth. **Plant and Soil,** Dordrecht, v.lll, p. 267-269, 1988.

BARBOSA, G. F.; CENTURION, M. A. P. C.; FERRAUDO, A. S. Potential of integrated management of Asian soybean rust: disease severity, vegetative development and yield components, cultivar MG/BR-46. **Bioscience Journal, Uberlândia,** v. 30, supplement 1, p. 76-89, June 2014.

BARETTA, D. et al. Performance of maize genotypes of Rio Grande do Sul using mixed models. **Científica,** v. 44, n. 3, p. 403, 18 Jul. 2016.

BARTMEYER, T. N.; DITTRICH, J. R.; SILVA, H. A.; MORAES, A.; PIAZZETTA, R. G.; GAZDA, T. L.; CARVALHO, P. C. F. Dual-purpose wheat submitted to cattle grazing in Campos Gerais, Paraná.

Pesquisa Agropecuária Brasileira, Brasília, v.46, n.10, p.1247-1253, 2011.

BASI, S.; NEUMANN, M.; MARAFON, F.; UENO, R.K.; SANDINI, I.E. Influence of nitrogen fertilisation on the quality of maize silage. **Pesquisa Aplicada & Agrotecnologia.** v4. n3. set/dez. (2011).

BASSO, C.J.; CERETTA, C.A. Nitrogen management in maize in succession to ground cover plants under no-tillage. **Revista Brasileira de Ciência do Solo,** v. 24, p. 905-915, 2000.

BASSOI, M. C.; BRUNETTA, D.; DOTTO, S. R.; SCHEEREN, P. L.Characteristics and agronomic performance in Paraná of the wheat cultivar BRS 220. **Pesquisa Agropecuária Brasileira,** v. 40, n. 2, p. 193-196, 2005.

BEAL, W. J. In: PATERNIANI, E.; VIEGAS, G. P. **Maize Improvement and Production.** Campinas: Cargill Foundation, 1987. 795p.

BECKER, H.C.; LÉON, J. Stabilityanalysis in plantbreeding. **Plant Breeding,** Berlin, v. 101, n. 1, p. 1-23,1988.

BELEZE, J.R.F.; ZEOULA, L.M.; CECATO, U.; DIAN, P.H.M.; MARTINS, E.N.; FALCÃO, A.J.S. Evaluation of Five Maize Hybrids (Zea mays, L.) at Different Ripening Stages. 1. Productivity, Morphological Characteristics and Correlations. **Revista Brasileira de Zootecnia,** v.32, n.3, p.529-537, 2003.

BENETT, C. G. S. et al. Foliar and top dressing application of nitrogen to wheat in the cerrado. **Semina: Ciências Agrárias,** Londrina, v. 32, n. 3, p. 829-838, jul/set. 2011.

BERLATO, M.A.; MATZENAUER, R. Teste de um modelo de estimativa do espigamento do milho com base na temperatura do ar, **Agronomia Sul Rio Grandense,** v. 22, p. 243-259,1986.

BERNARDI, I. P.; TEIXEIRA, E. M.; JACOMASSA, F. A. F. Relevant records of the avifauna of Alto Uruguai, Rio Grande do Sul, Brazil. **Biociências,** Porto Alegre, v. 16, n. 2, p. 134-137, 2008.

BERNARDO, R. **Breeding for quantitative traits in plants.** Woodbury: Stemma Press, 2002, 360p.

BLIGH, E.G.; DYER, W. J. A. Rapid method of total lipid extraction and purification. **Canadian Journal of Physiology and Biochemistry,** v.37, n.l,p.911-917,1959.

BLUM, B.L.E; AMARANTE, V.T.C; GÚTTLER, G.; MACEDO, F.A; KOTHE, M.D; SIMMLER, O.A; PRADO,G.; GUIMRÃES, S.L; Production of strawberry and cucumber in soil with incorporation of camasaviaria and pine bark. **Horticultura Brasileira,** v. 21, n.4, p. 627-631, 2003.

BOEHM, D. J.; BERZONSKY, W. A.; BHATTACHARYA, M. Influence of nitrogen fertiliser treatments on spring wheat (Triticumaestivum L.) flour characteristics and effect on fresh and frozen dough quality. **Cereal Chemistry,** St. Paul, v. 81, n. 1, p. 51-54,2004.

BOER, M. P. et al. A Mixed-Model Quantitative Trait Loci (QTL) Analysis for Multiple-Environment Trial Data Using Environmental Covariables for QTL-by-Environment Interactions, With an Example in Maize. **Genetics,** v. 177, n. 3, p. 1801-1813,1 nov. 2007.

BOREM, A.; MIRANDA, G. V. **Plant Breeding - 6th edition.** Viçosa: Editora UFV, 2013. v.l. 527p.

BORGES, V. et al. Genetic progress of the Minas Gerais highland rice improvement programme using mixed models. **Revista Brasileira de Biometria,** v. 27, n. 3, p. 478^490, 2009.

BORTOLINI, P. C.; SANDINI, L; CARVALHO, P. C. F.; MORAES, A. Winter cereals submitted to cutting in the dual-purpose system. **Revista Brasileira de Zootecnia,** Viçosa, v.33, n.l, p.45-50, 2004.

BORÉM, A. e MIRANDA, G.V. **Melhoramento de plantas.** 5 ed. Viçosa, MG. Ed: UFV, 2009. 529p.

BRADSHAW, A.D. Evolutionary significance of phenotypic plasticity in plants. **Adv. Gent.,** v.13, p. 115- 155, 1965.

BRAMMER, S.P.; MARTINELLI. P.; FERNANDES, de M.M.I.B.; PRESTES, A.M.; ANGRA, D.C. The potential of Agropyron, a species related to cultivated wheat, as a source for the introgression of agronomically important genes. **Embrapa Online Documents.** 2001.

BRANCOURT - HULMEL, M.; DOUSSINAULT, G.; LECOMTE, C.; BERARD, P.; LE BUANEC, B.; TROTTET, M. 2OO3.Genetic improvement of agronomic traits of winter wheat cultivars released in France from 1946 to 1992. Crop Breeding, Genetics & Cytology.Crop **Science,** v. 43, n. 1, p. 37.

BRANDT, E. A.; SOUZA L. C. F.; VITORINO, A. C. T.; MARCHETTI, M. E. Agronomic performance of soya beans as a function of crop succession in no-till farming. **Ciência & Agrotecnologia,** Lavras, v.30, n.5, p.869- 874, 2006.

BRAZIL, M. DA A. P. E A. 2009. **Rule for Seed Analysis.** Brasília. 399p.

BRAZIL. Ministry of Agriculture, Livestock and Supply. Normative Instruction 38 of 30 November 2010. **Technical regulations for wheat.** Official Gazette of the Federative Republic of Brazil, Brasília, Section 1, n.29, p.2, 1,2010.

BRAZIL. Ordinance No. 133 of 242 July 2014. **Approves Agricultural Climate**

Risk Zoning for the soya crop in the state of Rio Grande do Sul, crop year 2014/2015. Official Gazette of the Federative Republic of Brazil, Brasília, DF, 24 July 2014.

BRAZ, G. B. P.; CASSOL, G. M.; ORDONEZ, G. A. P.; SIMON. G. A.; PROCÓPIO, S. O.; OLIVEIRA NETO, A. M.; FERREIRA FILHO, W. C.; DAN, H. A. Production components and soya yield as a function of desiccation time and post-emergence management. **Revista Brasileira de Herbicidas,** Maringá, v. 9, n. 2, p. 63-72, 2010.

BREDEMEIER, C.; MUNDSTOCK, C. M. Phenological stages of wheat for nitrogen fertilisation. **Revista Brasileira de Ciência do Solo,** Porto Alegre, v.25, n.2, p.317-323, apr./jun. 2001.

BRENCHLEY, R.; SPANNGL, M,; PFEIFER, M.; BARKER, G. L. A.; AMORE, R. D.; ALLEN, A. M.; MCKENZIE, N.; KRAMER, M.; KERHORNOU, A.; BOLSER, D.; KAY, S.; WAITE, D.; TRICK, M.; BANCROFT, L; GU, Y.;HU, N,;CHENG LUO, N.; SEHGAL, S.; GILL, B.; KIANIAN, S.; ANDERSON, O.; KERSEY, P.; DVORAK, J.; MCCOMBIE, R.; HALL, A.; MAYER, M.; EDWARDS, K.; BEVAN, W.; HALL, H. Analysis of the bread wheat genome using whole-genome shotgun sequencing. **Nature,** London, v.491, 705-709, 2012.

BRUM, A.L.; HECK, C.R.; LEMES, C.L.; MÚLLER, P.K. 2005. The world soya economy: impacts on the oilseed production chain in Rio Grande do Sul 1970-2000. **In: XLIII SOBER CONGRESS,** Ribeirão Preto, SP.

BUCKLER, E. S.; STEVENS, N. M. Maize origins, domestication, and selection. In: MOTLEY, T. J.; ZEREGA, N.; CROSS, H. **Eds. New York: Columbia,** chap. 4, p. 67-90, 2006.

BUENO, L.G.; CHAVES, L.J.; OLIVEIRA, J.P.D.; BRASIL, E.M.; REIS, A.J.D.S. ASSUNÇÃO, A.; RAMOS, M. R. Genetic control of grain protein content and agronomic traits in maize grown with different levels of nitrogen fertilisation. **Pesquisa Agropecuária Brasileira,** v. 44. n. 6, 590-598, 2009.

BUENO, L. C. de S.; MENDES, A. N. G.; CARVALHO, S. P. **Plant Genetic Improvement: Principles and Procedures.** Editora UFLA, 2ª Ed. 2006. 319 p.

BUSSAB W. DE O. et al. 1990. **Introduction to cluster analysis.** São Paulo: Ed. Associação Brasileira de Estatística, lOOp.

BÚLL, L. T.; CANTARELLA, H. **Maize cultivation: Factors affecting productivity.** Piracicaba: POTAFOS, 1993. P.63-146.

CABEZAS, W.A.R.L.; SOUZA, M.A. Ammonia volatilisation, nitrogen leaching and maize productivity in response to the application of mixtures of urea with ammonium sulphate or with agricultural gypsum. **Revista Brasileira de Ciência do**

Solo, v. 32, n. 6, 2008, pp. 2331-2342.

CABRAL P. D. S. et al. 2011. DE. Trail analysis of bean grain yield (Phaseolus vulgaris L.) and its components. **Revista de Ciências Agronómica,** v. 42, p. 132-138, 2011.

CABRAL, L. S.; FILHO, S. C. V.; DETMANN, E.; ZERVOUDAKIS, J. T.; PEREIRA, O. G.; VELOSO, R. G.; PEREIRA, E. S. Rumen kinetics of carbohydrate fractions, gas production, in vitro digestibility of dry matter and estimated NDT of maize silage with different grain proportions. **Revista Brasileira de Zootecnia,** v.31, n.6, 2002.

CABRAL, L. S.; FILHO, S. C. V.; DETMANN, E.; ZERVOUDAKIS, J. T.; VELOSO, R. G.; NUNES, P. M. M. Digestion rates of protein and carbohydrate fractions for maize and elephant grass silages, Tifton-85 grass hay and soya meal. **Revista Brasileira de Zootecnia,** v.33, n.6, Viçosa-MG, 2004.

CAIRÃO, E.; SCHEEREN, P. L.; SILVA, M. S.; CASTRO, R, L. History of wheat cultivars released by Embrapa in forty years of research. **Crop Breeding and Applied Biotechnology,** Viçosa, v.14, n.l, p.216-223, 2014.

CALONEGO, J.C.; PALMA, H.N.; FOLONI, J.S.S. Foliar nitrogen fertilisation with ammonium sulphate and urea in maize cultivation. **Journa of Agronomic Sciences,** Umuarama, v.l, n.l, p.34-44, 2012.

CANCELLIER, L. L.; AFFÉRRI, F. S.; DUTRA, D. P.; LEÃO, F. F.; PELUZIO, J. M.; CARVALHO, E. D. Forage potential of maize populations in the south of the state of Tocantins. **Biosci. J.,** Uberlândia, v. 27, n. 1, p. 77-87, 2011.

CANCELLIER, L.L.; AFFÉRRI, F.S.; CARVALHO, E.V.; DOTTO, M.A.; LEÃO, F.F. Nitrogen use efficiency and phenotypic correlation in tropical maize populations in Tocantins. **Revista Ciência Agronómica,** v. 42, n. 1, p. 139-148, 2011.

CANTARELLA, H. Efficient use of nitrogen fertilisers: Efficient use of nitrogen in new fertilisers in Brazil. **Informações Agronómicas IPNI,** n. 120, p. 12-13, 2007.

CARGNELUTTI FILHO, A.; STORCK, L. Evaluation statistics of the experimental precision in com cultivar trials. **Pesquisa Agropecuária Brasileira,** v. 42, n. 1, p. 17-24, jan. 2007.

CARGNIN, A.; SOUZA, M. A. DE; CARNEIRO, P. C. S.; SOFIATTI, V. 2006. Interaction between genotypes and environments and implications for selection gains in wheat. **Pesquisa Agropecuária Brasileira,** v. 41, n. 6, p. 987- 993.

CARMO, M.S.do.; CRUZ, S.C.S.; SOUZA, E. J. de.; CAMPOS, L.F.C.; MACHADO, C.G. doses and sources of nitrogen in the development and productivity of the sweetcorn crop (Zeamaysconvar. saccharatavar. rugosa).

Bioscience Journal, Uberlândia, v. 28, Supplementl, p. 223-231, Mar. 2012.

CARPENTIERI-PÍPOLO V, SOUZA A, SILVA DA, BARRETO TP, GARBUGLIO DD, FERREIRA JM. Evaluation of Creole maize cultivars in a low-technology system. **Acta Scientiarum Agronomic,** v. 32, n. 2, p. 229-233, 2010.

CARVALHO I. R. et al. Water demand of crops of agronomic interest. **Enciclopédia Biosfera,** v. 9, p. 100-110, 2013.

CARVALHO, C. G. P.; ARIAS, C. A. A.; TOLEDO, J. F. F.; ALMEIDA, L. A.; KIIHL, R. A.; OLIVEIRA, M. F. Genotype x environment interaction in the productive performance of soya beans in Paraná. **Pesquisa Agropecuária Brasileira,** Brasília, v.37, n.7, p.989-1000, 2002.

CARVALHO, F. I. F.; LORENCETTI, C.; BENIN, G. **Estimates and implications of correlation in plant breeding.** Pelotas: UFPel, 142p. 2004.

CARVALHO, I. R. C.; SOUZA, V. Q. DE; NARDINO, M.; FOLLMANN, D. N.; SCHMIDT, D.; BARETTA, D. 2015. Canonical correlations between morphological characters and yield components in dual-purpose wheat. **Pesquisa Agropecuária Brasileira,** v. 50, n. 8, p. 690-697.

CARVALHO, I. R.; SOUZA, V. Q. de; NARDINO, M.; FOLLMANN, D. N.; DEMARI, G.; SCHMIDT, D.; SZARESKI, V. L; PELEGRIN, A. J. de; FERRARI, M.; PAVAN, M.; OLIVOTO, T. Effects of fungicides on soya beans with determinate growth habit. **Revista Sodebras,** v. 10, p.30- 34, 2015.

CARVALHO, I.R.; SOUZA, V.Q.; NARDINO, M.; FOLLMANN, D.N.; SCHMIDT, D.; BARETTA, D. Canonical correlations between morphological characters and yield components in dual-purpose wheat. **Pesquisa Agropecuária Brasileira,** v.50 n.8 Brasília, 2015.

CARVALHO, N. M.; NAKAGAWA, J. 2000. **Seeds: science, technology and production.** Jaboticabal: FUNEP.

CARVALHO, S.P. de. **Alternative methods for estimating path coefficients and selection indices under multicollinearity.** Viçosa: UFV, 1995.163p

CAZETTA, D.A.; FORNASIERI FILHO, D; ARF, O.; GERMANI, R. Industrial quality of wheat and triticale cultivars subjected to nitrogen fertilisation in the no-till system. **Bragantia,** Campinas, v. 67, n. 3, 2008.

CERETTA, C.A.; BASSO, C.J; FLECHA, A.M.T.; PAVINATO, P.S.; VIEIRA, F.C.B.; MAI, M.E.M. Management of nitrogen fertilisation in the black oat/maize succession in the no-till system. **Revista Brasileira de Ciência do Solo,** v. 26, p. 163-171, 2002.

CHANG, T. T. Availability of plant germplasm for use in crop improvement. In:

STALKER, H. T.; MURPHY, J. P. Plant breeding in the 1990s. **Melksham: Redwood Press,** 1992, p. 17-35.

CHAVARRIA, G.; DA ROSA, W. P.; HOFFMANN.L.; DURINGON, M. R.Growth regulator in wheat: effects on vegetative development, yield and grain quality. **Ceres,** v. 62, n. 6, 2015.

CIVARDI, E.A.; SILVEIRA NETO, A.N.; RAGAGNIN, V.A.; GODOY, E.R.; BROD, E. Slow-release urea applied superficially and common urea incorporated into the soil on maize yield. **Pesquisa Agropecuária Tropical,** v.41, n.01, p.52-59, 2011.

COELHO, A. M.; CRUZ, J. C.; PEREIRA FILHO, I. A. Challenges for obtaining high maize yields. In: Abstracts of the XXV National Maize and Sorghum Congress, Cuiabá, MT. **Sete Lagoas: ABMS/Embrapa Milho e Sorgo/Empaer,** 2004. p.186.

COELHO, A. M.; FRANCA, G. E. Be the doctor of your maize: nutrition and fertilisation. Informações Agronómicas, Piracicaba, n.71, Sep. 1995. **Arquivo do Agronómico, Piracicaba,** n. 2, p.1-9,1995.

COELHO, M.A.O.; SEDIYAMA, T.; SOUZA, M.A.; RIBEIRO, A.C.; SEDIYAMA, C.S. Mineral composition and nutrient export by grains of irrigated wheat subjected to increasing doses of nitrogen fertiliser. **Revista Ceres,** Viçosa, v. 48, n. 275, p.81-84, 2001.

COIMBRA, J. L. M.; BENIN, G.; VIEIRA, E.; OLIVEIRA, A. C.; CARVALHO, F. I. F.; GUIDOLIN, A. F.; SOARES, A. P. Consequences of multicollinearity on path analysis in canola. **Ciência Rural,** v. 35, n.02, p. 347-352, 2005.

COIMBRA, J. L. M.; GUIDOLIN, A. F.; CARVALHO, F. I. F. de; COIMBRA, S. M. M.; MARCHIORO, V. S. Trail analysis I: analysing grain yield and its components. **Ciência Rural,** Santa Maria, v. 29, n. 2, p. 213-218, 1999.

COIMBRA, J. L. M.; GUIDOLIN, A. F.; CARVALHO, F. I. F.; AZEVED, R. Canonical correlations: II - Analysing bean grain yield and its components. **Ciência Rural,** Santa Maria, v.30, n.l, p.31-35, 2000.

COIMBRA, J. L. M.;CARVALHO, F. I. F.; OLIVEIRA, C.; GUIDOLIN, A. F. Creation of genetic variability in the plant height character in oats: artificial hybridisation x induced mutation. **Revista Brasileira de Agrociência,** v. 10, p. 273-280, 2004.

COIMBRA, J.L.; CARVALHO, F.I.F.; OLIVEIRA, A.C.; DA SILVA, J.A.; LORENCETTI, C. Comparison between chemical and physical mutagens in oat populations. **Ciência Rural,** Santa Maria, v.35, n.l, p46-55, 2005.

COLLIER, L.S. KIKUCHI, F.Y.; BENÍCIO, L.P.F. SOUZA, S.A. de. Consortium and succession of maize and pigeonpea as an alternative crop under no-till farming.Pesquisa **Agropecuária Tropical,** vol.41 n.3 Goiânia July/Sept. 2011.

SOIL CHEMISTRY AND FERTILITY COMMISSION - RS/SC. **Fertilisation and liming manual for the states of Rio Grande do Sul and Santa Catarina.** lO.ed. Porto Alegre: SBCS - Southern Regional Centre/UFRGS, 2004. 400p.

CONAB - NATIONAL SUPPLY COMPANY. **Monitoring the Brazilian grain harvest,** v. 1 - 2013/2014 **Harvest,** n. 11 - Eleventh Survey, Brasília, p. 1-82, Aug. 2014.

CONAB - NATIONAL SUPPLY COMPANY. **Monitoring the Brazilian grain harvest,** v. 1 - 2014/2015 harvest. Brasilia: Agricultural Observatory, 2015. Pg. 1-103.

CONAB - National Supply Company. **Brazilian crop monitoring: grains, first survey, March 2014** / Companhia Nacional de Abastecimento. Brasília - DF: Conab, 2014.

CONAB - National Supply Company. **Brazilian crop monitoring: grains, first survey, March 2013** / Companhia Nacional de Abastecimento. Brasília: Conab, 2013.

CONAB-National Supply Company. **Monitoring the Brazilian Grain Harvest: Seventh survey/April 2016** - Brasília, v.3, n.7, p. 1-158, April 2016.

COSTA, M.C.G. Agronomic efficiency of nitrogen sources in sugar cane cultivation in a harvesting system without fire stripping. Master's dissertation. 79 p. Luiz de Queiroz College of Agriculture, Piracicaba, 2001.

COSTA, M.G. de SOUZA, E.L., STAMFORD, T.L.M.; ANDRADE, S.A.C. Technological quality of national and imported wheat grains and flours. **Revista de Ciência e Tecnologia de Alimentos,** v. 28, n. 1, p. 220-225, 2008.

COSTA, N. P.; MESQUITA, C. DE M.; FRANÇA-NETO, J. DE B.; MAURINA, A. C.; KRZYZANOWSKI, F. C.; OLIVEIRA, M. C. N. DE; HENNING, A. A. 2005.Validation of the ecological zoning of the state of Paraná for soya bean seed production. **Revista Brasileira de Sementes,** v. 27, n. 1, p. 37^44.

COVENTRY, D. R.; GUPTA, R. K.; YADAV, A.; POSWAL, R. S.; CHHOKAR, R. S.; SHARMA, R. K.; YADAV, V. K.; GILL, S. C.; KUMAR, A.; MEHTA, AKLEEMANN, S. G. L.; BONAMANO, A.; CUMMINS, J. A. 20íl.Wheatqualityandproductivity as affectedby variety andandsowing time in Haryana, india. **Field Crops Research,** v. 123, n. 3, p. 214-225.

CQFS, **Manual de Adubação e de Calagem. Soil Chemistry and Fertility Commission** - RS/SC, 394p., Porto Alegre, 2004.

CRESPO, A.A. **Estatística Fácil.** 17ed. São Paulo: Saraiva, 2002.

CRUZ C. D. & REGAZZI, A. J. **Biometric models applied to genetic improvement.** Viçosa: 2 Ed UFV, 450p.

CRUZ C. D. et al. 2012. **Biometric Models Applied to Genetic Improvement.** Viçosa: 4 Ed UFV, 400p.

CRUZ, C. D. Genes - a software package for analysis in experimental statistics and quantitative genetics. **Acta Scientiarum Agronomy.** Maringá, v.35, p.271-276, 2013.

CRUZ, C. D.; CARNEIRO, P. C. S. **Modelos biométricos aplicados ao melhoramento genético,** v.2, 2.ed. Viçosa: Ed. UFV, 2006. 585p.

CRUZ, C. D.; CARNEIRO, P. C. S.; REGAZZI, A. J. **Biometric models applied to genetic improvement.** Viçosa: Editora UFV, v.2, n.3, 2014, 668 p.

CRUZ, C. D.; REGAZZI, A. J. **Biometric models applied to genetic improvement.** 2 ed. Viçosa: UFV, 1997. 390p.

CRUZ, C. D.; REGAZZI, A. J. **Biometric models applied to genetic improvement.** 2. ed. Viçosa: UFV, 2001.

CRUZ, C. D.; REGAZZI, A. J. **Biometric models applied to genetic improvement.** Viçosa: UFV, 1994. 380 p.

CRUZ, C. D.; REGAZZI, A. J.; CARNEIRO, P. C. S. **Biometric models applied to genetic improvement.** 2.ed.Viçosa: Ed. da UFV, 2012. 514p.

CRUZ, C. D.; REGAZZI, A. J.; CARNEIRO, P. C. S. **Biometric models applied to genetic improvement.** 3. Ed. Viçosa, MG: UFV, 2004. v. 1,480 p.

CRUZ, C.D. **Genes Programme - Experimental Statistics and Matrices.** Ed. 2. Editora UFV. Viçosa, 2006. 285p.

CRUZ, J. C. Maize cultivars for silage. In: CONGRESSO NACIONAL DOS ESTUDANTS DE ZOOTECNIA, 1998, Viçosa, MG. **Proceedings... Viçosa: UFV,** 1998. p. 93-114.

CRUZ, J.C.; FILHO, I.A.P.; GARCIA, J.C.; DUARTE, J. O. **Cultivo do Milho. Embrapa Maize and Sorghum.** Sistemas de Produção, 2 Versão Eletrónica -6ª edition. 2010.

CRUZ, J.C.; PEREIRA-FILHO, I.A.; NETO, M.M.G. Maize for Silage. **Embrapa Technological Information Agency,** 2016.

CRUZ, S.C.S.; PEREIRA, F.R. da S.; SANTOS, J.R.; ALBUQUERQUE de, A.W.; PEREIRA, R.G. Nitrogen fertilisation for maize grown in a no-till system in the state of Alagoas. **Revista Brasileira de Engenharia Agrícola e Ambiental,** v.12, n.l, p.62-68, 2008.

CUNHA, C.A.H. **Relationship between spectral behaviour, leaf area index and dry matter production in Tanzania grass subjected to different irrigation levels and nitrogen doses.** Thesis (Doctorate in Agronomy) - Luiz de Queiroz College of Agriculture, Piracicaba, 2004. 154p.

CÂMARA TMM, BENTO DAV, ALVES GF, SANTOS MF, MOREIRA JUV, SOUZA JÚNIOR CD. Genetic parameters of characters related to tolerance to water deficiency in tropical maize. **Bragantia** v. 66, n. 4, 595-603, 2007.

DALCHIAVON, F.C.; CARVALHO, M.P. 2012. Linear and spatial correlation of production components and soya productivity. **Semina: Ciências Agrárias,** Londrina, v.33, p.541-552.

DE ARAÚJO AV, JÚNIOR DDSB, FERREIRA ICPV, DA COSTA CA, PORTO BBA. Agronomic performance of creole varieties and maize hybrids grown in different management systems. **Revista Ciência Agronómica,** v. 44, n. 4, p. 885-892, 2013.

DE JESUS FREITAS, I. L. et al. Genetic gain assessed with selection indices and REML/Blup in popcorn. **Pesquisa Agropecuária Brasileira,** Brasília, v. 48, n. 11, p. 1464-1471, 2013.

DE RESENDE, M. D. V. et al. Best linear unbiased prediction (BLUP) of genetic values in Pinus improvement. **Forestry Research Bulletin,** 1996.

DE SOUZA, V. Q. et al. Potential of selection among and within potato clonal families. **Crop Breeding and Applied Biotechnology,** v. 5, n. 2, p. 199-206, 2005.

DEDECCA, D. M.; PURCHIO, M. J. Agricultural varieties of wheat (Triticum aestivumL.), **Bragantia,** Campinas, v.12, n.3, p. 19-53, 1952.

DEL DUCA, L. J. A.; GUAIENTI, E. M.; FONTANELI, R. S.; ZANOTTO, D. L. Influence of cuts simulating grazing on the chemical composition of winter cereal grains. **Revista Agropecuária Brasileira,** Brasília, v.34, n.9, p. 1607-1614,1999.

DEL DUCA, L. J. A.; MOLIN, R.; SANDINI, I. Trialling dual-purpose wheat genotypes in Paraná in 1999. **Passo Fundo- Embrapa Wheat - Research and Development Bulletin,** v. 6, 2000, 18p.

DEMARI G. H. et al.. Non-preferential season beans subjected to nitrogen doses and their impact on agronomic characters. **Enciclopédia Biosfera,** v. 11, p. 1102-1112, 2015.

DERERA, N. Preharvest field sprouting in cereals. **Boca Raton: CRC Press,** p. 176. 1989.

DOVALE, J. C. et al. Genetic effects of traits associated with nitrogen use efficiency in maize. **Pesquisa Agropecuária Brasileira,** v. 47, n. 3, p. 385-392, 2 Aug. 2013.

DUETE, R.R.C.; MURAOKA, T.; SILVA, E.C.; AMBROSANO, E.J.; TRIVELIN, P.C.O. Accumulation of nitrogen (15n) by maize grains as a function of nitrogen source in red latosol (1). **Bragantia,** Campinas, v.68, n.2, p.463-472, 2009.

DUVICK DN. The contribution of breeding to yield advances in maize (Zea mays L.). **Advances in agronomy,** San Diego 86: 83-145, 2005.

EAST, E. M. HETEROSIS. Genetics, v. 24, p. 375-397. 1936.

EAST, E. M. The distinction between development and heredity in breeding. **American Naturalist,** v. 43, p. 173-181,1909.

EBERHART, S. A.; SALHUANA, W.; SEVILLA, R.; TABA, S. Principies for tropical maize breeding. **Maydica,** Bergamo, v. 40, p. 339- 355, 1995.

EGLI, D.B.; MECKEL, R.E.; PHILLIPS, R.E.; RADCLIFFE, D.; LEGGETT, J.E. 1983. Moisture stress and N redistribution in soybean. **Agronomy Journal,** Madison, v.75, p.1027-1031.

EMBRAPA - Brazilian Agricultural Research Corporation. Brazilian soil classification system. **Brasília: EMBRAPA,** 2006. 306p,

EMBRAPA - Brazilian Agricultural Research Corporation. Soya production technologies - central Brazil 2009 and 2010. **Londrina: Embrapa Soja: Embrapa Cerrados: Embrapa Agropecuária Oeste,** 2008. 262 p.

BRAZILIAN AGRICULTURAL RESEARCH COMPANY - EMBRAPA. National Soybean Research Centre. Technical recommendations for soya cultivation in central Brazil 1996/97, **Londrina: EMBRAPA - CNPSO,** 1996.

149p.ESCOSTEGUY, P. A. V.; RIZZARDI, M. A.; ARGENTA, G. Doses and timing of nitrogen topdressing on maize in two sowing seasons. **Revista Brasileira de Ciência do Solo,** Campinas, v. 21, n. 1, p. 71-77, 1997.

EURIDES, L.P.; AGUIAR, A.P.A.; BONILHA, M.A.F.M.; RAFAEL, H.M.; CASETA, M.C. Influence of nitrogen fertilisation with different nitrogen sources in the form of ammonium sulphate, ammonium nitrate and urea. **FAZU em Revista,** Uberaba, n. 5, p.89-93, 2008.

FALCONER D. S. **Introduction to quantitative genetics.** Viçosa: UFV, 1997. 200p.

FALCONER, D. S. **Introduction to quantitative genetics.** Viçosa: UFV, 4 ed. 1996. 464p

FANCELLI, A.L.; DOURADO-NETO, D. **Corn Production.** Ed. Agropecuária, Guaíba. 360 p., 2000.

FARIA MV, MENDES MC, ROSSI ES, JÚNIOR OP, RIZZARDI DA, GRALAK E, FARIA CMDR. Diallel analysis of yield and progress of leaf disease severity in maize hybrids at two population densities. **Semina: Agricultural Sciences,** v36, n. 1, p. 123- 134, 2015.

FEDERIZZI, L. C.; SCHEREN, P. L.; NETO, J. F. B; MILACH, S. C. K P.; PACHECO, M. T. Wheat improvement. In: Borém. A Improvement of cultivated species. Viçosa: UFV, 1999, 535-587p.

FEHR, W. R. Principies of cultivar development. **New York: MacMillan,** 1987. 525p.

FELINI, F.Z; BONO, J.A.M. Productivity of Soya and Maize in a Planting System with the Use of Chicken Manure in the Region of Sidrolândia - MS. Trials and Science. **Agricultural, Biological and Health Sciences,** v. 15. N.5, 2011.

FERNANDES, F.C.S.; BUZETTI, S.; ARF, O.; ANDRADE, J.A.C. Doses, efficiency and use of nitrogen by six maize cultivars. **Revista Brasileira de Milho e Sorgo,** v.4, n.2, p. 195-204, 2005.

FERREIRA, J.M.; MOREIRA, R.M.P.; HIDALGO, J.A.F. Combinatorial capacity and heterosis in creole maize populations. **Ciência Rural,** v. 39, n.2, p. 332-339, 2009.

FERREIRA, A.C.B.; ARAÚJO, G.A.A.; PEREIRA, P.R.G.; CARDOSO, A.A. Agronomic and nutritional characteristics of maize fertilised with nitrogen, molybdenum and zinc. **Scientia Agrícola,** v.58, n.l, p.131-138, jan./mar. 2001.

FERREIRA, F. M.; BARROS, W. S.; SILVA, F. L.; BARBOSA, M. H. P.; CRUZ, C. D. Phenotypic and genotypic relationships between yield components in sugarcane. **Bragantia,** Campinas, v. 66, n. 4, p. 605-610, 2007.

FILHO, A. N. K.; DE RESENDE, M. D. V. Scientific notes on variance components and prediction of genetic values in rubber trees using the mixed modelling methodology (reml/blup). **Pesquisa Agropecuria Brasileira,** Brasília, v. 35, n. 9, p. 1883-1887, 2000.

FILHO, M.C.M.T.; BUZETTI, S.; ANDREOTTI, M., ARF, O.; BENETT, C.G.S.B. Doses, sources and times of nitrogen application in irrigated no-till wheat. **Pesquisa agropecuária brasileira,** Brasília, v.45, n.8, p.797-804, August 2010.

FIOREZE, S. L. and RODRIGUES, J. D.2014.Wheat production components affected by sowing density and plant regulator application. **Semina: Ciências Agrárias,** vol. 35, n. 1, p.39-54.

FONTANELI, R. S.; FONTANELI, R. S.; SANTOS, H. P.; JÚNIOR, A. N.; MINELLA, E.; CAIRÃO, E. Yield and nutritional value of dual-purpose winter cereals: green forage and silage or grain. **Revista Brasileira de Zootecnia,** Viçosa, v.38, n.ll, p.2116-2120, 2009.

FONTANELI, R. S.; SANTOS, H. P.; FONTANELI, R. S. Forage crops for crop-livestock-forest integration in the South of Brazil. 2Br. Ed. Brasília, **DF: Embrapa Brasília,** 2012. 544p.

FONTANELI, R. Trigo duplo propósito na integração lavoura-pecuária, **Revista Plantio Direto,** Passo Fundo, v.99, n.l, p. 10-20, 2007.

FONTOURA, T. B.; COSTA, J. A.; DAROS, E. Effects of levels and times of defoliation on the yield and grain yield components of soya beans. **Scientia agrária,** v.7, n.l, p. 49-54, 2006.

FORNASIERI Filho, D. **Maize cultivation.** Jaboticabal: FUNEB, 1992. 273p.

FRANK, A. B.; BAUER, A. Temperature, nitrogen, and carbon dioxide effects on spring wheat development and spikelet numbers. **Crop Science,** v. 36, n. 3, p. 659-665,1996.

FRANÇA, S.; MIELNICZUK, J.; ROSA, L.M.G.; BERGAMASCHI, H.; BERGONCI, J.I. Nitrogen available to maize: Growth, absorption and grain yield. **R. Bras. Eng. Agríc. Ambiental,** v.15, n.ll, p.1143- 1151,2011.

FUERTES-MENDIZÁBAL, T.; AIZPURUA, A.; GONZÁLEZ-MORO, M.B.; ESTAVILLO, J.M. Improving wheat bread-making quality by splitting the N fertilizer rate, **European Journal of Agronomy.V.** 33, Issue 1, p. 52-61, July 2010

FUZATTO, S. R.; FERREIRA, D. F.; RAMALHO, M. A. P.; RIBEIRO, P. H. E. Genetic divergence and its relationship with diallel crosses in maize. **Ciência e Agrotecnologia,** Lavras, v. 26, n. 1, p. 22-32, 2002.

GARCIA, A.; PÍPOLO, A. E.; LOPES, I. de O. N.; PORTUGAL, F. A. F. Soya planting: season, cultivars, spacing and plant population. **Technical circular, Embrapa, Londrina,** PR. September 2007.

GILL, B. S.; FRIEBE, B.; ENDO, T. R. Standard karyoty peand nomenclature 17 system for description of chromosome bands and structural aberrations 18 in wheat (Triticum aestivum). **Genome,** v.34, n.l, 830-839. 1991.

GOMES JÚNIOR, F. G. et al. 2005. Protein content in bean grains at different times and doses of nitrogen top dressing. Acta Scientiarum. Agronomy 27, 100-110.

GOMES, M.S.; PINHO, R.G.V.; RAMALHOM, M.A.P.; FERREIRA, D.F. Alternatives for selection of maize hybrids involving several characters for silage production. **Revista Brasileira de Milho e Sorgo,** v.5, n.3, p.406-421, 2006.

GOMES, M.S.; PINHO, R.G.V.; RAMALHOM, M.A.P.; FERREIRA, D.V.; BRITO, A.H. Genetic variability in maize lines for characteristics related to silage yield. **Pesquisa agropecuária brasileira,** Brasília, v.39, n.9, p.879-885, sep. 2004.

GONZALEZ, S.; CÓRDOVA, H.; RODRIGUEZ, S.; DE LEON, H.; SERRATO, V. M. Determinación de un patrón heterotico a partir de la evaluacion de undialelo de diezlineas de maiz subtropical. **Agronomia Mesoamericana,** v. 8, p. 1-7, 1997.

GRALAK, E.; FARIA, M. V.; JÚNIOR, O. M.; ROSSI, E. S.; SILVA, C. A.; RIZZARDI, D. A.; MENDES, M. C.; NEUMANN, M. Combinatorial capacity of maize hybrids for agronomic and bromatological characters of silage. **Revista Brasileira de Milho e Sorgo,** v.13, n.2, p. 187-200, 2014.

GRIFFING, B. Concept of general and specific ability in relation to diallel crossing systems. **Australian Journal of Biological Sciences,** v. 9, n. 4, p. 462-493, 1956.

GUARIENTI, E.M.; DE BONA, F.D.; PIRES, J.L.F.; NICOLAU, M.; STRIEDER, M.L.; SCHEEREN, P. L.; WIETHÔLTER, S. Nitrogen and the technological quality of wheat. **Technical note - Embrapa Wheat.** 2013.

GUBIANI, É.I. 2005. **Growth and yield of soya beans in response to sowing times and plant arrangement,** 62p. Dissertation (Master's in Plant Science) - Federal University of Rio Grande do Sul.

GULLUOGLU, L.; ARIOGLU, H., KURT, C. Adaptability and stability of new soybean cultivars under double cropped conditions of Turkey. **African Journal of Agriculture! Research,** Lagos, v. 6, n. 14, p. 3320-3325, 2011.

HALLAUER, A. R. History, Contribution, and Future of Quantitative Genetics in Plant Breeding: Lessons From Maize. **Crop Science,** v. 47, n. 3, p. 4-19, 2007.

HALLAUER, A. R.; CARENA, J. M.; MIRANDA FILHO, J. B. Quantitative genetics in maize breeding. New York: Springer, 2010. 500p.

HAN, G. C.; VASAL, S. K.; BECK, D. L.; ELIAS, E. Combining ability of inbred lines derived from CIMMYT maize (Zea mays L.) germplasm. **Maydica,** v. 36, p. 57-64,1991.

HARPER, J.E. Nitrogen metabolism. In: BOOTE, KJ. et. al. Physiology and determination of crop yield. **American Society of Agronomy,** 1994. Chap.llA, P.285-302.

HARTWIG, L; CARVALHO, F. I. F.; OLIVEIRA, A. C.; VIEIRA, E. A.; SILVA, J. A. G.; BERTAN, L; RIBEIRO, G.; FINATTO, T.; REIS, C. E. S.; BUS ATO, C. C.; Estimation of correlation coefficient and track in segregating generations of hexaploid wheat. **Bragantia,** Campinas, v.66, n.2, p.202-218, 2007.

HASTENPFLUG, M.; MARTIN, N.; CASSOL, L. C.; BRAIDA, J. A.; BARBOSA, D. K.; MOCHINSKI, A.; Vegetative performance of dual-purpose wheat cultivars subjected to nitrogen fertilisation. **Revista FZVA,** Uruguaiana, v.16, n.l, p. 66-78. 2009.

HAYMAN, B. I. The theory and analysis of diallel crosses. **Genetics,** v. 39, p. 789-809, 1954.

HEIFFING, L.S. 2002. **Plasticity of the soya bean crop (Glycine max (L.) Merrill) in different spatial arrangements,** 85p. Dissertation (Master's Degree in Agronomy) - University of São Paulo/School of Agriculture "Luiz de Queiroz".

HEINRICHS, R.; AITA,C.; AMADO, T.J.C.; FANCELLI, A.L. Intercropping oats and vetch: C/N ratio of phytomass and yield of maize in succession. **Revista Brasileira de Ciência do Solo,** 25:331- 340, 2001.

HIROMOTO, D.M.; CAMACHO, S.A. INOX Soya: More productivity and resistance to rust. MT Foundation in Focus. Year 05 - No. 25 - June/July 2008 **(MT Foundation Bimonthly Newsletter).**

HUBBARD, M.; GERMI; VUJANOVIC, V. 2O12.Fungal endophytes improve wheat seed germination under heat and drought stress. **Botany,** v. 90, n. 2, p. 137- 149.

RIO GRANDE DO ARROZ INSTITUTE - IRGA (2014). Survey of the area sown with soya in lowlands in Rio Grande do Sul, 2014. Porto Alegre: IRGA. Retrieved 10 December 2015, from http://www.irga.rs.gov.br.

JESUS WC, BRASIL EM, OLIVEIRA JP, PINTO GRC, CHAVES LJ, RAMOS MR. Heterosis for grain protein content in crosses between maize populations derived from commercial hybrids. **Pesquisa Agropecuária Tropical,** v. 38, n. 1, p. 32-38, 2008.

JONES, D. F. The effect of inbreeding and crossbreeding upon development. Genetics. **Connecticut Agricultural Experiment Station,** 207, p. 246-250,1918.

JÚNIOR, F.B.R.; MACHADO, C.T.T.; MACHADO, A.T.; SODEK,L. Inoculation of azospirillum amazonense in two maize genotypes under different nitrogen regimes. **Revista Brasileira de Ciência do Solo,** v. 32, p. 1139-1146, 2008.

JÚNIOR, G. L. M.; ZANINE, A. D.; BORGES, L; PÉREZ, J. R. O. Fibre quality for ruminant diets. **Ciência Animal,** v. 17, n. 1, p. 7-17, 2007.

KARNOPP, Loiva et al. Cytological effects of the pyrethroid insecticide deltamethrin on barley (Hordeumvulgare L.). CurrentAgricultural **Science and Technology,** v. 5, n. 2,1999.

KEMPTHORNE, O.; CURNOW, R. N. The partial diallel cross. **Biometrics.** v. 17, p. 50 - 229.1961.

KIEHL, J.C. Distribution and retention of ammonia in soil after urea application. **Revista Brasileira de Ciência do Solo,** v. 13, n. 1, p. 75-80,1989.

KNAPP, S. J.; STROUP, W. W.; ROSS, M. W. Exact confidence intervals for heritability on a progeny mean basis. **Crop Science,** Madison, v. 25, p. 192-194, 1985.

KOESTER RP, SISCO PH, STUBER CW. Identification of Quantitative Trait Loci Controlling Days to Flowering and Plant Height in Two Near Isogenics Lines of Maize. **Crop Science,** Madison 33(6): 1209-1216, 1993.

KRUG, C. A.; VIÉGAS, G. P.; PAOLIÉRI, L. Commercial maize hybrids. **Bragantia,** v. 3, p. 367-552, 1943.

KU, L. X.; ZHAO, W. M.; ZHANQ, J.; WU, L. C.; WANG, C. L.; WANG, P. A.; ZHANG, W. Q; CHEN, Y. H. Quantitative trait loci mapping of leaf angle and leaf orientation value in maize (Zea mays L.). **Theoretical and Applied Genetics,** v. 121, p. 951-959, 2010.

KUCHEL, H.; LANGRIDGE, P.; MOSIONEK, L.; WILLIAMS, K.; JEFFERIES, S.P. The genetic control of milling yield, dough rheology and baking quality of wheat. **Theoretical and Applied Genetics,** Berlin, v.112, n.8, p.1487-1495, 2006.

KUREK A. et al. 2012. Path analysis as an indirect selection criterion for grain yield in beans. **Current Agricultural Science and Technology,** v. 7, p. 25-35.

KUREK, A. L; CARVALHO, I. F. de; ASSAMANN, I. C.; MARCHIORO, V. S.; CRUZ, P. J. Path analysis as an indirect selection criterion for grain yield in beans. **Revista Brasileira de Agrociência,** v.7 n.l, p. 29-32, 2001.

LAMOTHE, Adriana Garcia. **N fertilisation and yield potential in wheat. Exploring high wheat yields,** p. 209, 1998.

LARGE, E. C. Growth stages in cereals illustration of the Feekes scale. **Plant pathology,** v. 3, n. 4, p. 128-129, 1954.

LEMAIRE, G.; GASTAL, F. N. N uptake and distribution in plantcanopies. **In: LEMAIRE, G. (Ed.). Diagnosis of the nitrogen status in crops.** Berlin: Springer, 1997. p. 3-43.

LI, C. C. Path analysis - a primer. Boxwood: Pacific Grove, 1975. 346p.

LIMA J. L.; DE SOUZA, C. L; MACHADO, J. C.; RAMALHO, M. A. P. Genetic control of the thermal requirement for the onset of flowering in maize. **Bragantia,** Campinas, n. 67, v.l, p. 127-131, 2008.

LIMA, E. V.; CRUSCIOL, C. A. C.; CAVARIANI, C.; NAKAGAWA, J. Agronomic characteristics, productivity and physiological quality of "safrinha" soya beans under direct sowing, as a function of mulching and surface liming. **Revista Brasileira de Sementes,** v. 31, n. 1, p. 69-80, 2009.

LOPES UV, GALVÃO JD, CRUZ CD. Inheritance of the flowering time in maize. **Pesquisa Agropecuária Brasileira,** v. 30, n. 10, 1267- 1271, 1995.

MAC RITCHIE, F.; GUPTA, R.B. Functionality composition relationships of wheat flour as a result of variation in sulphur availability. **Australian Journal of**

Agriculture! Research, Victoria, v.44, n.8, p. 1767-1774,1993. MACHADO, A.T.; MAGALHÃES, F.R.; MAGNAVACA, R.; SILVA, M.R e. Determination of the activity of enzymes involved in nitrogen metabolism in different maize genotypes. **Revista Brasileira Fisiologia vegetal,** v. 14, n. 1, p. 45-47,1992.

MAEHLER, A. R; COSTA, J. A; PIRES, J. L; RAMBO. L. Grain quality of two soya cultivars as a function of soil water availability and plant arrangement. **Ciência Rural,** Santa Maria, v.33, n.2, p.213- 218, 2003.

MAGALHÃES, P. C.; DURÃES, F. O. M. Maize cultivation, germination and emergence. **Technical Communiqué 39,** Ministry of Agriculture, Livestock and Supply, Sete Lagoas, MG, 2002.

MAGALHÃES, P.C.; DURÃES, F.O.M.; CARNEIRO, N.P.;PAIVA,E. Maize Physiology. **Technical Circular (22),** p.23. Sete Lagoas, MG, 2003.

MAHLER, Robert L.; KOEHLER, Fred E.; LUTCHER, L. K. Nitrogen source, timing of application, and placement: effects on winter wheat production. **Agronomy Journal,** v. 86, n. 4, p. 637-642, 1994.

MALAVOLTA, E. **Nitrogen fertilisers.** In: ABC da adubação. São Paulo: Agronómica Ceres, 1989. P. 26-34.

MALAVOLTA, E. **Elementos da nutrição mineral de plantas.** São Paulo: Agronómica Ceres, p. 251, 1981

MARCOS FILHO, J. 1999. Vigour tests: importance and use. **In: Seed vigour: concepts** and tests.Londrina: ABRATES, p. 1-21.

MARQUES, B. M. F. P. P.; ROSA, G. B.; HAUSCHILD, L.; CARVALHO, A.d'A.; LOVATTO, P. A. Replacement of maize by low tannin sorghum in diets for pigs: digestibility and metabolism. **Arquivo Brasileiro de Medicina Veterinária e Zootecnia,** Belo Horizonte, v.59, n.3, p.767-772, 2007.

MARQUES, M.C. 2014. **Performance of crosses between genitors tolerant to Asian soya bean rust,** 357p. Thesis (Doctorate in Science: Genetics and Plant Breeding). University of São Paulo/School of Agriculture "Luiz de Queiroz".

MARTIN, T. N.; SIMINATTO, C. C.; ORTIZ, P. B. S.; HASTENPFLUG, M.; ZIECH, M. F.; SOARES, A. B. Phytomorphology and yield of dual-purpose wheat cultivars in different cutting managements and sowing densities. **Ciência Rural,** Santa Maria, v.40, n.8, p.1695- 1701, 2010.

MARTIN, T. N.; STORCK, L.; BENIN, G.; SIMIONATTO, C. C.; ORTIZ, S.; BERTOCELLI, P. Importance of the relationship between characters in dual-purpose wheat. **Bioscience Journal,** Uberlândia, v.29, n.6, p.1932- 1940, 2013.

MARY, B.; RECOUS, S.; DARWIS, D.; ROBIN, D. Interactionbetween decomposition of plant residues and nitrogen andnitrogen cycling in soil. **Plant Soil,** The Hague, 181:71-82,1996.

MAUAD, M.; SILVA, T. L. B.; NETO, A. I. A.; ABREU, V. G. Influence of sowing density on agronomic characteristics of the soya bean crop. **Revista Agrarian,** Dourados, v.3, n.9, p. 175-181, 2010.

MEDEIROS, S. L. P.; WESTHPHALEN, S. L.; MATZENAUER, R.; BERGAMASCHI, H. Relationships between evapotranspiration and maize grain yield. **Pesquisa Agropecuária Brasileira,** Brasília, v.26, n.l, p.1-10, 1991.

MEDEIROS, S.R.; ALBERTINI, T.Z.; MARINO, C.T. Lipids in ruminant nutrition. **In: MEDEIROS, S.R.; GOMES, R.C.; BUNGENSTAB, D.J. Beef cattle nutrition: fundamentals and applications.** Brasília, DF: Embrapa, 2015.

MEDINA, P.F.; RAZERA, L.F.; MARCOS FILHO, L; BORTOLETTO, N. 1997.Seed production of early soya cultivars in two seasons and two locations in São Paulo: II. Physiological quality, **Bragantia,** v. 56, n.2, p.305-315.

MEINERZ, G. R.; OLIVO, c. J.; VIÉGAS, J.; NORNBERG. J. L.; AGNOLIN, A.; SCHEIBLER, R. B.; HORST, T.; FONTANELI, R. S. Winter cereal silage submitted to dual-purpose management. **Revista Brasileira de Zootecnia,** Viçosa, v.40, n.10, p.2097-2104,2011.

MEIRA, F.A.; BUZETTI, S. **Nitrogen sources and application methods in maize cultivation.** Doctoral dissertation, São Paulo State University, Ilha Solteira Campus. Ilha Solteira - SP December/2006.

MEIRA, F.de A.; BUZETTI, S. In: DUARTE, A.P.; PATERNIANI, M.; ANDREOTTI, M.; ARF, O.; SÁ, M.E de.; ANDRADE, J.A. da C. Sources and times of nitrogen application in irrigated maize. **Semina: Ciências Agrárias,** Londrina, v. 30, n. 2, p. 275-284, Apr./Jun. 2009.

MELERO, M. M., DE CASTILHO GITTI, D., ARF, O., & RODRIGUES, R. A. F.2013. Vegetable covers and nitrogen doses in wheat under no-till system. **Pesquisa Agropecuária Tropical,** Goiânia,p 343- 353.

MELO, R.; NORNBERG, J. L.; ROCHA, M. G. Productive and qualitative potential of maize, sorghum and sunflower hybrids for silage. **Revista Brasileira de Agrociência,** v. 10, n. 1, p. 87-95, 2004.

MENDOZA D. F. **Nitrogen losses by ammonia volatilisation and efficiency of nitrogen fertilisation in irrigated rice.** Dissertation (master's degree) - Federal University of Santa Maria, Centre for Rural Sciences, Postgraduate Programme in Soil Science, RS, 85p. 2006. MESQUITA, L.A.V. Ammonium nitrate. Agronomic Information n° 120, 6-7p - December/2007.

MENEGALDO, J. G. The importance of maize in people's lives. **In: Grupo**

Cultivar de **Publicações** Ltda. Pelotas, 2013.

MENEZES, J. F. S.; ANDRADE, C. de L. T.; ALVARENGA, R. C.; KONZEN, E. A.;PIMENTA, F. F. Chicken litter in agriculture: perspectives and technical and economic viability. **Rio Verde: FESURV,** 2004. 28 p. (Technical bulletin, 3).

MENGEL, D.B.; BARBER, S.A. Rate of nutrient uptake per unit of com root under field conditions. **Agronomy Journal,** v.66, p.399-402, 1974.

MERCHING, J.S.; DOWLER, W.; KOOGLE, D.L.; ROYER, M.H. 1989. Effects of duration, frequency, and temperature of leaf wetness periods on soybean rust. **Plant Disease,** Saint Paul, v.73, p. 117-122.

MINGOTTE F. L. C. et al. 2014. Predecessor cropping systems and nitrogen topdressing doses in no-till bean cultivation. **Bioscience Journal,** v. 30, p. 100-110.

MIRANDA FILHO, J. B.; GORGULHO, E. P. Crosses with testers and diallels. In: NASS, L. L.; VALOIS, A. C. C.; MELO, I. S.; VALADARES, M. C. (Ed.) Genetic resources and plant breeding. **Rondonópolis: MT Foundation,** 2001. p. 650-671.

MIRANDA, G. V.; COIMBRA, R. R.; GODOY, C. L.; SOUZA, L. V.; GUIMARÃES, L. J. M.; MELO, A. V. Improvement potential and genetic divergence of popcorn cultivars. **Pesquisa agropecuária brasileira.** Brasília, v. 38, n. 6, p. 681-688, June 2003.

MIRANDA, M.Z.; GUARIENTI, E.M. TONON, V.D. Technological quality of wheat. In: PIRES, J. L. F.; VARGAS, L.; CUNHA, G. R. da (ed.). Wheat in Brazil: Bases for competitive and sustainable production. **Passo Fundo: Embrapa Trigo,** 2011. p. 371-389.

MITTELMANN, A.; NETO, J. F. B.; CARVALHO, F. I. F.; LEMOS, M. C. I.; CONCEIÇÃO, L. D. H. Inheritance of wheat characters related to baking quality. **Pesquisa Agropecuária Brasileira,** Brasília, v.35, n.5, p.975-983, 2000.

MITTELMANN, A.; SOBRINHO, F.S.; OLIVEIRA, J.S.; FERNANDES, S.B.V; LAJÚS, C.A.; MIRANDA, M.; ZANATTA, J.C.; MOLETTA, J.L. Evaluation of commercial maize hybrids for use as silage in the Southern Region of Brazil. **Ciência Rural,** Santa Maria, v.35, n.3, p.684-690, mai-jun, 2005.

MOMEN, N.N.; CARLSON, R.E.; SHAW, R.H.; ARJMAND, O. 1979. Moisture-stress effects on the yield components of two soybean genotypes. **Agronomy Journal,** Madison, v.71, p.86-90.

MONTGOMERY, E.G. Correlation studies of com. Annual report. **Agricultural Experiment Station,** Nebraska. v.l, p. 108-159. 1911.

MORAES FERNANDES, M. I. B. Cytogenetics, wheat in Brazil. **Campinas: Cargill Foundation,** 1982, 95-144p.

MORAES, L.B.D. de; FREO, J.D.; BIDUSKI, B.; ELIAS, M.C.; GUTKOSKI, L.C. Effects of Rate, Time and Splitting of Nitrogen Fertilisation on the Technological Quality of Wheat. **Journal of Food Science and Engineering** 3p. 9-18, 2013.

MORAES, S. D.; JOBIM, C. C.; SILVA, M. S.; MARQUARDT, F. I. Production and chemical composition of sorghum and maize hybrids for silage. **Revista brasileira saúde de produção animal,** vol.14 no.4 Salvador Oct./Dec. 2013.

MOREIRA, A.L.; PEREIRA, O.G.; GARCIA, R.; FILHO, S.C.V.; CAMPOS, J.M.S; MORAES, S.A.; ZERVOUDAKIS, J.T. Consumption and Apparent Digestibility of Nutrients from Maize Silage and Alfalfa and Coastcross Grass Hays in Sheep. **Revista brasileira de zootecnia,** v. 30, n. 3, p. 1099-1105, 2001.

MORENO, José Alberto. 1961. **Climate of Rio Grande do Sul.** Porto Alegre, Department of Agriculture, 42p.

MORRISON, D. F. **Multivariate statistical methods.** 2. Ed. Tokyo: McGraw HUI, 1978,415p.

MOTA, F.S. da. Study of the climate of the state of Rio Grande do Sul, according to W. Koeppen's system. **Revista Agronómica,** Porto Alegre, v.8, n.193, p.132-141,1953.

MOTURI, B.; CHARYA, M. S. Influence of physical and Chemical mutagens ondyede colourising Mucormucedo. **African Journal of Microbiology** Research, v. 4, n. 17, p. 1808-1813, 2010.

MUNDSTOCK, C., THOMAS, A. L. Soya: Factors affecting growth and grain yield. Porto Alegre: Department of Crop Plants, Federal University of Rio Grande do Sul, **Evangraf,** 2005. 31p.

MUNDSTOCK, C.M. **Planning and integrated management of wheat crops.** Porto Alegre: Evnagraf, 1999. 227p.

McBLAIN, B.A.; HUME, D.J. 1981. Reproductive abortion yield components and nitrogen content in threeearly soybean cultivars. **Canadian Journal of Plant Science,** v.61, p.499-505.

McINNES, K.J & FILLERY, I.R.P. Modelling and field measurements of the effect of nitrogen source on nitrification. **Soil Sei. Soc. Am. J.,** v. 53, p. 1264-1269,1989'.

MÓDENES, A.N.; SILVA, A. .; TRIGUEIROS, D.E.G. Evaluation of the rheological properties of stored wheat. **Ciênc. Tecnol. Aliment.,** Campinas, v. 29, n. 3, p. 508-512, 2009.

MÚLLER, M. M. L.; CECCON, G.; ROSOLEM, C. A. Influence of subsurface soil compaction on aerial and root growth of winter green manure plants. **Revista Brasileira de Ciência do Solo,** v. 25, n. 3, p. 531-538, sep. 2001.

NARDINO, M. et al. Association of secondary traits with yield in maize F 1 's. Ciência Rural, v. 46, n. 5, p. 776-782, May 2016.

NASCIMENTO, M.; FINOTO, E. L.; SEDIYAMA, T.; CRUZ, C.D. Adaptability and stability of soybean in teim so foil and protein content. Crop Breeding and Applied Biotechnology, Viçosa, v. 10, n. 1, p. 48-54, 2010.

NATIONAL RESEARCH COUNCIL - NRC. Nutrient requirements of dairy cattle. 7th rev. ed. Washinton, D.C.: 2001. 38Ip.

NAVARRO JÚNIOR, H. M.; COSTA, A. C. Relative contribution of growth components to soya bean production. Pesquisa Agropecuária Brasileira, Brasília, v. 37, n. 2, p. 269-274, 2002.

NEUMANN, M.; MÚHLBACH, P.R.F.; RESTLE, J. et al. Maize silage (Zea mays, 1.) at different cutting heights and particle size: production, composition and utilisation in feedlot cattle finishing. Revista Brasileira de Milho e Sorgo, v.6, n.3, p.379-397, 2007.

NOGUEIRA A.R.A.; G.B. SOUZA. Laboratory manual: soil, water, plant nutrition, animal nutrition and food. São Carlos: Embrapa Pecuária Sudeste, v.l, p. 313. 2005.

NOGUEIRA, A. P. O.; SEDIYAMA,T.; SOUSA, L. B.; HAMAWAKI, O. T.; CRUZ, C. D.; PEREIRA, D. G.; MATSUO, E. Trail analysis and correlations between characters in soya beans grown at two sowing times. **Bioscience Journal,** Uberlândia, v.28, n.6, p.877-888, 2012.

NOVAIS, R.F.; ALVARES, V.H.V.; BARROS, N.F. de.; FONTES, R.L.F.; CANTARUTTI, R.B.; NEVES, J.C.L. **Fertilidade do Solo.** Brazilian Society of Soil Science. Iª edition, 1017 p, Viçosa, Minas Gerais, 2007.

NOVINSKI, C.O.; SOUZA, C.M. de; SCHMIDT, P. Bromatological characterisation of maize silages in Brazil. **Forage Research Centre - UFPR, 2013.**

NUNES, E.N.; MONTENEGRO, I.N.A.; NASCIMENTO, D.A.M; SILVA, D.A.; NASCIMENTO, R. Growth analysis and nitrogen assimilation in maize plants (Zeamays L.). **Revista Verde (Mossoró - RN - Brasil),** v. 8, n. 4, p. 72 - 76, oct - dec, 2013.

NUSSIO, L. G. Production of high quality silage. In: CONGRESSO NACIONAL DE MILHO E SORGO, 19, 1992, Porto Alegre. **Conferences...Porto Alegre: SAA/SCT/ABMS/Emater-RS/Embrapa- CNPMS,** 1992. p. 155-175.

OKUYAMA, L. A.; FERERIZZI, L. C.; BARBOSA NETO, J. F. Correlation and path analysis of yield and its components and plant traits in wheat. **Ciência Rural,** Santa Maria, v. 34, n. 6, p. 1701-1708, 2004.

OLIBONI, R.; FARIA, M. V.; NEUMANN, M.; RESENDE, J. T. V.; BATTISTELLI, G. M.; TEGONI, R. G.; OLIBONI, D. F. Dialectical analysis in the evaluation of the potential of maize hybrids for the generation of base populations to obtain lines. **Semina: Ciências Agrárias,** Londrina, v. 34, n. 1, p. 7-18, jan./feb. 2013.

OLIVEIRA, A. C.; COIMBRA, J.; CARVALHO, F.; GUIDOLIN, A. Creation of genetic variability in the plant height character in oat: artificial hybridisation x induced mutation. Current **Agricultural Science and Technology,** v. 10, n. 3, 2012.

OLIVEIRA, L.B.; PIRES, A.J.V.; CARVALHO, G.G.P.; RIBEIRO, L.S.O.; ALMEIDA, V.V.; PEIXOTO, C.A.M. Losses and nutritional value of maize, sorghum-sudan, forage sorghum and sunflower silages. **Revista Brasileira de Zootecnia,** v.39, pp.61-67.2010.

OVENTRY, D.; REEVES, T.; BROOKE, H.; CANN, D. 1993.1nfluence of genotype, sowing date, and seeding rate on wheat development and yield. **Australian Journal of Experimental Agriculture,** v. 33, n. 6, p. 751-757.

OZTURK, A.; CAGLAR, O.; BULUT, S. Growth and yield response of facultative wheat to winter sowing, freezing sowing and spring sowing at different seeding rates. **Crop Science,** Madison, v.192, n.4, p.10-16, 2006.

PAINI JN, CRUZ CD, DELBONI JS, SCAPIM CA. Combinatorial capacity and heterosis in maize inter-varietal crosses evaluated under the climatic conditions of southern Brazil. **Ceres,** v. 43, n. 247, 288-300, 2015

PANDEY, J.P.; TORRIE, J.H. 1973. Path coefficient analysis of seed yield components in soybean Glycine max (L.) Merrill. **Crop Science,** Madison, v.13, p.505-507.

PANDOLFO, C. M. **Technical, economic and environmental aspects of the use of organic nutrient sources associated with soil preparation systems.** PhD thesis. UFSM. 2005.

PATERNIANI, E. **Recent studies on heterosis.** Bulletin n°1. Cargill Foundation, São Paulo, 1974. 36p.

PATERNIANI, E. **Maize improvement and production in Brazil.** Campinas: Cargill Foundation, 1978. 650 p.

PATERNIANI, E.; CAMPOS, M. S. **Maize improvement.** In: BORÉM, A. (ed.). Improvement of cultivated species. Viçosa: UFV, 1999. 429-486 p.

PATERNIANI, E.; VIEGAS, G. P. Maize **Improvement and Production.** Campinas: Cargill Foundation, 1987. 795p.

PATERNIANI, M. E. A. G. Z. Use of heterosis in maize breeding: History, Methods and Perspectives. **Crop Breeding and Applied Biotechnology,** v. 1,n. 2p.159-178, 2001.

PATTERSON, H. D.; THOMPSON, R. Biometrika. **Recovery of inter- block Information when block sizes are unequal,** v. 58, p. 545-554, 1971.

PAZIANI, S. de F.; DUARTE, A.P.; NUSSIO, L.G.; GALLO, P.B.; BITTAR, C.M.M.; ZOPOLLATTO, M.; RECO, P.C. Agronomic and bromatological characteristics of maize hybrids for silage production. **Revista Brasileira de Zootecnia,** v.38, n.3, p.411-417,2009.

PEIXOTO, C. P.; SOUSA CÂMARA, G.M. de,; MARTINS, M. C.; MARCHIORI, L. F. S.; GUERZONI, R. A.; MATTIAZZI, P. Sowing times and plant density for soya: I. Production components and grain yield. **Scientia Agrícola,** v. 57, n.l, Piracicaba, 2000.

PELTONEN, Jari. Ear developmental stage used for timing supplemental nitrogen application to spring wheat. **Crop Science,** v. 32, n. 4, p. 1029- 1033, 1992.

PEREIRA, E. S., DE ARRDA, A. M. V., MIZUBUTI, I. Y., MOREIRA, F. B., CAVALCANTE, M. A., DE OLIVEIRA, S. M. P., & VILLARROEL, A. B. S. (2007). Bromatological composition and rumen kinetics of neutral detergent fibre of maize and sorghum silages. **Semina: Ciências Agrárias,** v. 28, n. 3, 529-534.

PEREIRA, F.B.; VALE do, J.C.; CARNEIRO, P.C.S.; NETO, R.F. Relationship between characters determining nitrogen and phosphorus use efficiencies in maize. **Revista Ceres,** vol.60 n.5 Viçosa Sept./Oct. 2013.

PERUZZO, G. Nitrogen in your wheat. **Cultivar Grandes Culturas magazine,** n.16, 2000.

PETTINELLI NETO, A.; CRUSCIOL, A.C.; BICUDO, S.J.; FREITAS, J.G. & PULZ, A.L. Efficiency and response of irrigated wheat genotypes to nitrogen for the State of São Paulo. In: CONGRESSO DE INICIAÇÃO CIENTIFICA, 14, Presidente Prudente, 2002. **Proceedings...** Presidente Prudente, UNESP-Scientific Initiation Programme, 2002.

PIMENTEL, A. J. B.; SOUZA, M. A.; CARNEIRO, P. C. S.; ROCHA, J. R. A. S. C.; MACHADO, J. C.; RIBEIRO, G. Partial diallel analysis in advanced generations for selection of wheat segregating populations, **Pesquisa Agropecuária Brasileira,** Brasília, v.48, n.12, p.1555-1561, 2013.

PIPOLO, V. C.; GASTALDI, L. F.; PIPOLO, A. E. Phenotypic correlations between quantitative traits in soya beans. **Semina: Ciências Agrárias,** Londrina, v. 26, n. 1, p. 11-16, 2005.

PIRES, A.J.V.; REIS, R.A.; CARVALHO, G.G.P.; SIQUEIRA, G.R.; BERNARDES, T.F.; RUGGIERI, A.C.; ROTH, M.T.P. Rumen degradability of dry matter, crude protein and fibre fraction of maize, sorghum and Brachiaria brizantha silages. **Arquivo Brasileiro de Medicina Veterinária e Zootecnia,** v.62, n.2, p.391-400, 2010.

PRA, M.A.D.; KONZEN, E.A.; OLIVEIRA, P.A.; MORES, E.; Composting Liquid Swine Manure. Embrapa.Documentos 45.SeteLagoas, MG. December 2005.

QUEIROZ, A.M; SOUZA, C.H; MACHADO, V.J; LANA, R.M.Q, KORNDORFER, G.H; SILVA, A.A. Evaluation of different sources and doses of nitrogen in the fertilisation of maize (Zeamays L.). **Revista Brasileira de Milho e Sorgo,** v.10, n.3, p. 257-266, 2011.

RAGAGNIN, V.A.; JÚNIOR, D.G.S.; KLEIN, V.; LIMA, R.S.; COSTA, M.M.; NETO, O.V.O. Nitrogen fertilisation on off-season com crop under notillage system in Jataí-GO. **Global Science and Technology.**

RAGAGNIN, V.A; JÚNIOR, de S. D.G.; DIAS, D.S; BRAGA, W.F; NOGUEIRA, M.P.D. Development and nodulation of soya plants fertilised with poultry litter. **Ciência Agrotecnologia,** vol.37 no.l Lavras. 2013.

RAMALHO M, SANTOS JB, PINTO CB, SOUZA EA, GONÇALVES FMA, SOUZA JC. **Genetics in Agriculture.** 5ed. Lavras: UFLA 565p, 2012.

RAMALHO, M. A. P.; ABREU, A. F. B.; SANTOS, J. B.; NUNES, J. A. **R. Applications of quantitative genetics in the improvement of autogamous plants.** Lavras: UFLA, 2012. 522p.

RAMALHO, M. A. P.; LAMBERT, E. S. Biometrics and plant breeding in the genomics era. **Revista Brasileira de Milho e Sorgo,** v. 3, n. 2, p. 228-249, 2004.

RAMALHO, M. A. P.; SANTOS, J. P. dos; ZIMMERMANN, M. J. de O. **Quantitative genetics in autogamous plants: Applications to bean improvement.** Goiânia: ED. da UFG. 1993. 27 Ip.

RAMALHO, M. SANTOS, J. B. PINTO, C. B. **Genética na Agropecuária.** 2° Edition. Editora Globo S. A., 1990, 359p.

RAMALHO, M.; SANTOS, J. B.; PINTO, C. B.; SOUZA, E. A.; GONÇALVES, F. M. A.; SOUZA, J. **C. Genética na Agropecuária.** 5 ed. Lavras: UFLA, 2012. 565p.

RAMBO, L.; COSTA, J. A.; PIRES, J. L. F.; PARCIANELLO, G.; FERREIRA, F. G. Yield potential estimated by soybean canopy stratum in response to plant angement. **Ciência Rural,** Santa Maria, v.34, n.l, p.33- 40, 2004.

RANGEL, O.J.P.; SILVA, C.A. Carbon and nitrogen stocks and organic fractions of latosol submitted to different use and management systems. **Revista Brasileira de Ciência do Solo,** 31:1609-1623, 2007.

REICHARDT, K.; TIMM, L.C. **Soil, Plant and Atmosphere. Concepts, processes**

and applications. 2 ed.500p. Barueri, SP. 2012.

REIS, E. F.; REIS, M. S.; CRUZ, C. D.; SEDIYAMA, T. Expected response to correlated selection in an F6 soya population. **Revista Ceres,** Viçosa, v. 48, n. 276, p.169- 179, 2001.

RESENDE, M. D. V. Software Selegem - REML/BLUP: statistical system and computerised genetic selection via linear mixed models. **Colombo: Embrapa Florestas,** 2007.

RESENDE, M. D. et al. Estimation of genetic parameters and prediction of genotypic values in coffee plant breeding using the REML/BLUP procedure. **Bragantia,** v. 60, n. 3, p. 185-193, 2001.

REZENDE, P. M.; CARVALHO, E. de A. Evaluation of soya cultivars [Glycine max (L.) Merrill] for the south of Minas Gerais. **Ciência e Agrotecnologia,** Lavras, v.31, n.6, p.1616-1623, 2007.

RIBEIRO, A. M. L.; HENN, J. D.; SILVA, G. L. Alternative feeds for growing and finishing pigs. **Acta Scientia e Veterinariae,** Porto Alegre, v.38, n.l, p.61- 71, 2010.

RIBEIRO, G.; PIMENTE, A. J.; SOUZA, M. A.; ROCHA, J. R. A. S. C.; FONSECA, W. B. High temperature stress in wheat: impact on development and tolerance mechanisms. **Revista Brasileira Agrociência,** Pelotas, v.18, n.2-4, p.133-142, 2012.

ROBERTSON, J.B.; VAN SOEST P.I. The detergent system of analysis and its application to human foods. **The analysis of dietary fibre in food,** p.123-158. 1981.

RODRIGUES, O. Wheat management: ecophysiological bases. In: Cunha, G. R.; Bacaltchuk, B. Tecnologia para produzir trigo no Rio Grande do Sul. **Porto Alegre: Legislative Assembly of Rio Grande do Sul,** 2000. p. 120-169. Crop Series - Wheat.

ROIAS, B. A.; SPRAGUE, G. F. A comparison of variance components in com yield trials. III. General and specific combining ability and their interaction with location and years. **Agronomy Journal,** Madison, v. 44, p. 462-466, 1952.

ROSA FILHO, O. Introduction to Management for Industrial Quality in Wheat. **Biotrigo. Technical Bulletin 1/2010,** 6p.

ROSA, R.P.R.; SILVA, J.H.S; RESTLE, 1; PASCOAL. L.L.; BRONDANI, I.L.; FILHO, D.C.A.; FREITAS, A.K. Evaluation of Plant Agronomic Behaviour and Nutritive Value of Silage from Different Maize Hybrids (Zea mays, L.). **Revista Brasileira de Zootecnia,** v.33, n.2, p.302-312, 2004.

ROSADO, A. M. et al. Simultaneous selection of eucalyptus clones according to productivity, stability and adaptability. **Pesquisa Agropecuária Brasileira,** v. 47, n. 7, p. 964-971, 2012.

ROSTAGNO, H.S; ALBINO, L.F.T.; DONZELE, I.L.; GOMES, P.C. **Tabelas Brasileiras para aves e suínos: Composição de alimentos e exigências nutricionais.** Federal University of Viçosa. Department of Zootechnics. 3. ed. - Viçosa, MG, 2011,254p.

SAGGAR, A.S et al. 14C-labelled ryegrass tumover and residence times in soils varying in clay content and mineralogy. **Soil Biology & Biochemistry,** v.28, p. 1677-1686, 1996.

SALMAN, A.K.D.; FERREIRA, A.C.D.; SOARES, J.PG.; SOUZA, J.P. Metodologias para avaliação de alimentos para ruminantes domésticos. Porto Velho, **RO: Embrapa Rondônia,** 2010. 21 p.

SANCHEZ, E. **Soil physical properties and soya productivity in succession to winter cover crops.** Master's Thesis. Guarapuava, State University of the Centre-West, 2012,48p.

SANGOI L, ALMEIDA ML, SILVA PRF, ARGENTA G. Morphological basis for greater tolerance of modern maize hybrids to high plant densities. **Bragantia,** Campinas, v. 6, n. 2, p. 101-110, 2012.

SANGOI, L. et al.,. Early nitrogen top dressing does not increase grain yield of wheat grown in the presence of aluminium. **Ciência Rural,** v. 38, n. 4, p. 912-920, 2008.

SANGOI, L.; BERNS, A. C.; ALMEIDA, M. L.; ZANIN, C. G.; SCHWEITZER, C. Agronomic characteristics of wheat cultivars in response to top dressing nitrogen fertilisation. **Ciência Rural,** Santa Maria, v.37, n.6, p.1564-1570, 2007.

SANGOI, L.; GUIDOLIN, A. F.; COIMBRA, J. L. M; SILVA, P. R. F. Response of maize hybrids cultivated at different times to plant population and to peeling. **Ciência Rural,** Santa Maria, v. 36, n. 5, p. 1367-1373, Sep/Oct 2006.

SANGOI, L.; VARGAS, V.P.; SCHIMITT, A.; PLETSCH, A.J.; VIEIRA, J.; SALDANHA, A.; SIEGA, E.; CARNIEL, G.; MENGARDA, R.T.; JÚNIOR, G.J.P. Nitrogen availability, survival and contribution of tillers to maize grain yield. **Revista Brasileira de Ciência do Solo.** v. 35, p. 183-191,2011.

SANGOI, L.;ERNANI, P.R.; BIANCHET, P. Initial development of maize as a function of doses and sources of nitrogen applied at sowing. **Biotemas,** v. 22, n. 4,53-58, 2009.

SANTOS, H. G.; JACOMINE, P. K. T.; ANJOS, L. H. C.; OLIVEIRA, V. A.; OLIVEIRA, J. B.; COELHO, M. R.; LUMBRERAS, J. F.; CUNHA, T. J. F.; **Brazilian soil classification system.** Rio de Janeiro: Embrapa Solos, 2006. 306 p.

SANTOS, H. P.; FONTANELI, R. S.; CAIRÃO, E.; SPERA, S. T.; VARGAS. L.. Agronomic performance of wheat grown for grain and dual purpose in crop-livestock integration systems. **Pesquisa agropecuária brasileira,** Brasília, v.46, n.10, p.1206-1213, 2011.

SANTOS, J.; VENCOVSKY, R. Phenotypic and genetic correlation between some agronomic characters of the bean plant (Phaseolus vulgaris L.). **Ciência e Prática,** Lavras, v.10, n.3. p.265-272, 1986.

SANTOS, P.A.; SILVA, A.F.; CARVALHO, M.A.C.; CAIONE, G. Green manures and nitrogen fertilisation in maize cultivation. **Revista Brasileira de Milho e Sorgo,** v.9, n.2, p.123-134, 2010.

SANTOS, R. C.; CARVALHO, L. P.; SANTOS, V. F. Analysis of path coefficients for peanut production components. **Ciência e Agrotecnologia,** Lavras, v.24, n.l, p.13-16, 2000.

SAS INSTITUTE. **Statistical Analysis System, SAS/STAT Useris Guide 9.3,** North Caroline, NC: SAS, 2012.

SCALCO, M. S. et al. Productivity and industrial quality of wheat under different levels of irrigation and fertilisation. **Ciência e Agrotecnologia,** v. 26, n. 2, p. 400-410, 2002.

SCHEEREN, P. L.; CAIERÃO, E.; SILVA, M. S.; BONOW, S. Wheat improvement in Brazil In: Trigo no Brasil. Wheat improvement in Brazil. l^a ed. **Passo Fundo: Embrapa Trigo,** v. 01, n.l, 2011, 427- 452p.

SCHERER, E. E.; BALDISSERA, I. T. Using pig waste as fertiliser. In: FIELD DAY ON THE MANAGEMENT AND USE OF PIG WASTE, Concórdia. **Proceedings...** Concórdia: Embrapa Swine and Poultry: EPAGRI: FATMA, 1994. p. 33-37.

SCHEUER, P.M.; FRANCISCO, A.; MIRANDA, M.Z.; LIMBERG, V.M. Wheat: characteristics and use in baking. **Revista Brasileira de Produtos Agroindustriais,** Campina Grande, v. 13, n. 2, p. 211-222, 2011.

SCHIAVINATTI, A.F; ANDREOTTI, M; BENETT, C.G.S; PARIZ, C.M; LODO, B.N; BUZETTI,S. Influence of nitrogen sources and application methods on production components and productivity of irrigated maize in the cerrado. **Bragantia,** Campinas, v. 70, n. 4, p.925-930, 2011.

SCHMIDT, D. A. M.; CARVALHO, F. I. F.; OLIVEIRA, A. C.; SILVA, J. A. G.; BERTAN, I; VALÉRIO, I. P.; HARTWIG, I; SILVEIRA, G.;

GUTKOSKI, L. C. Genetic variability in Brazilian wheat from component characters of industrial quality and grain yield. **Bragantia,** Campinas, v.68, n.l, p.43-52, 2009...

SCIVITTARO, W.B.; MURAOKA, T.; BOARETTO, A.E.; TRIVELIN, P.C.O.

Utilisation of nitrogen from green and mineral manures by maize. **Revista Brasileira de Ciência do Solo,** v.24, p.917-926, 2000.

SEARLE, S. R.; CASELLA, G.; MCCULLOCH, C. E. **Variance components.** 2. ed. New York: J. Willey, 1992.

SEDIYAMA, T. **Production technologies and uses of soya.** Londrina: Macenas, 2009.

SEDIYAMA, T.; PEREIRA, M. G.; SEDIYAMA, C. S.; GOMES, J. L. L. **Soya cultivation.** Viçosa: UFV, 1996. 96p.

SEMA - STATE SECRETARIAT FOR THE ENVIRONMENT. **Turvo State Park management plan.** Porto Alegre, 2005. p. 348.

SENA JÚNIOR, D.G.; PINTO, F. de A. de C.; QUEIROZ, D.M. de; SANTOS, N.T.; KHOURY, J.K. Discrimination between nutritional stages in the wheat crop with artificial vision techniques and portable chlorophyll meter. **Engenharia Agrícola,** v.28, n.l, p.187-195, 2008.

SENGER, C. C. D.; KOZLOSKI, G. V.; SANCHEZ, L. M. B.; MESQUITA, F. R.; ALVES, T. P.; CASTAGNINO, D. S. Evaluation of autoclave procedures for libre analysis in forage and concentrate feedstuffs. Castagnino **Animal Feed Science and Technology,** v.l, n.l, p. 100-120, 2008.

SHAPIRO, S.S; WILK, M.B. Analysis of variance test for normality (complete samples). **Biometrika,** p. 591-611,1965.

SHULL, G. H. A pure-line method of com breeding. Am. **Breeders Assoe. Rep.** v. 5, p. 51-59,1909.

SHULL, G. H. Hybrid Seed Com. **Science,** v. 103, n. 2679, p. 547-550, 1946.

SIAL, M. A.; ARAIN, M. A.; KHANZADA, S.; NAQVI, M. H.; DAHOT, M. U.; NIZAMANI, N. A. 2OO5.Yield and quality parameters of wheat genotypes as affected by sowing dates and high temperature stress. **Pakistan Journal of Botany,** v. 37, n. 3, p. 575.

SILVA H. 2005. Minimum descriptors for characterising cultivars/varieties of common bean (Phaseolus vulgaris L.). **Brasília: Ed Embrapa,** 25p.

SILVA, A.A.; SILVA, T.S.; VASCONCELOS, A.C.P. de; LANA, R.M.Q. Application of different sources of gradual release urea to maize. **Bioscience Journal,** Uberlândia, v. 28, supl., p. 104-111, 2012

SILVA, A.V.; PEREIRA, O.G.; GARCIA, R. et al. Bromatological composition and in vitro dry matter digestibility of maize and sorghum silages treated with microbial inoculants. **Revista Brasileira de Zootecnia ,** v.34, n.6, p.1881-1890, 2005.

SILVA, C.A.; VALE, F.R. Nitrate availability in Brazilian soils under the effect of liming and nitrogen sources. **Pesquisa Agropecuária Brasileira,** 35. P. 2461-2471, 2000.

SILVA, D. J.; QUEIROZ, C. **Food analysis: chemical and biological methods.** Federal University of Viçosa, 2006, l0Op.

SILVA, E.C.; BUZETTI, S.; GUIMARÃES, G.L.; LAZARINI, E.; SÁ, E. Doses and times of application of nitrogen to the maize crop in no-till on Red Latosol. **Revista Brasileira Ciência solo,** vol.29 no.3 Viçosa May/June 2005..

SILVA, M.G.O.; FREITAS, F.C.L.; MESQUITA, H.C.; NASCIMENTO, P.G.M.L.; RODRIGUES, A.P.M.S.; SANTANA, F.A.O. Grain yield of maize cultivars in consortium with Brachiariabrizantha. ACSA - **Agropecuária Científica no Semi-Árido,** v.7, n. 01 January/March 2011 p. 23 - 29.

SILVA, P.R.F. da et al. Grain yieldandkemel protein content increases of maizeh ybridswith late nitrogenside-dresses. **Scientia Agrícola,** Piracicaba, v.62, p.487-492, 2005.

SILVA, R. S.; BENIN, G.; SILVA, G. O.; MARCHIORO, V. S.; ALMEIDA, J. L.; MATEI, G. Adaptability and stability of wheat cultivars at different sowing times in Paraná. **Pesquisa Agropecuária Brasileira,** vol. 46, n.ll, p.1439-1447,2011.

SILVA, S. A.; CARVALHO, F. I. F. de; NEDEL, J. L.; CRUZ, P. L; PESKE, S. T.; SIMIONI, D.; CARGNIN, A. Seed filling in quasi-isogenic wheat lines with presence and absence of the "stay-green" character. **Pesquisa Agropecuária Brasileira,** Brasília, v.38, n.5, p.613-618, 2003.

SILVA, S. A.; CARVALHO, F.; COSTA, F.; COIMBRA, J.; LORENCETTI, C. Effect of the mutagens sodium azide and ethyl methane sulphonate, in the ml generation, in wheat (Triticum aestivum L.). **Current Agricultural Science and Technology,** v. 4, n. 2,1998.

SILVA, S. A.; DE CARVALHO, F, I. F.; NEDEL, J. L.; CRUZ, P. L; PESKE, S. T.; SIMIONI. D.; CARGNIN. A. Seed filling in near-isogenic wheat lines with presence and absence of the "stay-green" character. **Pesquisa Agropecuária Brasileira,** v. 38, n. 5, p. 613-618, 2003

SILVA, T.R.; MENEZES, J.F.S.; SIMON, G.A.; ASSIS, R.L.; SANTOS, C.J.L.; GOMES, G.V.; Maize cultivation and P availability under chicken litter fertilisation. **Rev. bras. eng. agríc. ambient.** vol.15 no.9 Campina Grande Sept. 2011.

SIMMONDS, I; SCOTT, P.; LEVERINGTON-WAITE, M.; TURNER, A.; BRINTON, J.; KORZUN, V.; SNAPE, L; UAUY, C. Identification and independent validation of as tableyieldandthousandgrain weight QTL on chromosome 6A of hexaploid wheat (Triticumaestivum L.).Bmc **Plant Biology,** v. 14, 2014.

SIMMONS,S. R.; RASMUSSSON, D. C.; WIERSMA, J. V. Tillering in barley:

genotype, row spacing and seeding rate effects. **Crop Science,** Madison, v. 22, n. 4, p. 801- 805, 1982.

SINGH, D. The relative importance of characters affecting genetic divergence. The **Indian Journal of Genetic and Plant Breeding,** v. 41, p. 237-245, 1981.

SINGH, R.; NYE, P.H.The effect of soil pH and high urea concentrations on urease activity in soil. **Journal of Soil Science,** v. 35, n. 4, p. 519-527, 1984.

SINGH, S.P. 2001. Broadening the genetic base of common bean cultivars: a review. **Crop Science,** Stanford, v.41, p. 1659-1675.

SIONTT, N.; KRAMER, PJ. 1977. Effect of water stress during different stages of growth of soybeans. **Agronomy Journal, Madison,** v.69, p.274- 278.

SKOVMAND, B.; RAJARAM, S.; RIBAUT, J.M.; HEDE, A. R. Wheat genetic resources. **Bread wheat: improvement and production.** Rome: FAO, 2002. p.89-102.

SLAGEREN, M. W. van. **Wild wheats: a monograph of Aegilops L. and Amblyopyrum (Jaub. & Spach) Eig (Poaceae).** Wageningen: Wageningen University, 1994. 513 p.

SNIFFEN, C.J.; O'CONNOR D.J.; VAN SOEST, P.J.; FOX, D.G.; RUSSELL, J.B.A net carbohydrate and protein system for evaluating cattle diets: carbohydrate and protein availability. **Journal of Animal Science,** v.l, p.3562-3577,1992.

SOARES SOBRINHO, J. **Effect of nitrogen doses and water levels on the agronomic and industrial characteristics of two wheat cultivars (Triticumaestivum L.).** 102p. Thesis (Doctorate in Plant Production) - Faculty of Agricultural and Veterinary Sciences, Paulista State University, Jaboticabal, 1999.

SOKAL R. R. & ROHLF, F. J. 1962. The Comparison of Dendrograms by Objective Methods. **Taxon,** 11, 33^40.

SORATTO, R.P; COSTA, T.A.M; FERNANDES, A.M; PEREIRA, M; MARUYAMA, W.I. Splitting of alternative nitrogen sources in maize in succession to soya. **Revista de Ciências Agrarias.**v.40, n.2, p. 179-188, 2012.

SOUSA, D.M.G. de; LOBATO, E. Liming and fertilisation for annual and semi-annual crops. In: SOUSA, D.M.G. de; LOBATO, G. Cerrado: correção do solo e adubação. **2. ed. Brasília: Embrapa Informação Tecnológica,** p.283-315, 2004.

SOUTO, P.C.; SOUTO, J.S.; SANTOS, R.V.; ARAÚJO, G.T.; SOUTO, L.S. Decomposition of manures at different depths in degraded areas in the semi-arid region of Paraiba. **Revista Brasileira de Ciência do Solo,** v. 29, p. 125-130, 2005

SOUZA LV, MIRANDA GV, GALVÃO JCC, ECKERT FR, MANTOVANI ÉE, LIMA RO, GUIMARÃES LJM. Genetic control of grain yield and nitrogen use efficiency in tropical maize. **Pesquisa Agropecuária Brasileira,** v. 43, n. 11, p. 1517-1523, 2008.

SOUZA, A. R. R.; MIRANDA, G. V.; PEREIRA, M. G.; FERREIRA, P. L. Correlation of characters in a Creole maize population for a traditional cultivation system. **Caatinga,** Mossoró, v.21, n.4, p.183-190, 2008.

SOUZA, C. A.; FIGUEREDO, B. P.; COELHO, C. M. M.; CASA, R. T.; SANGOI, L. Plant architecture and soya productivity resulting from the use of growth reducers. **Bioscience Journal,** Uberlândia, v. 29, n. 3, p. 634-643, 2013.

SOUZA, D. S. M; POLIZEL, A. C; HAMAWAKI, O. T.; BONFIM- SILVA, E. M.; KOETZ, M.; HAMAWAKI, R. L.. Selection of semi-early/medium cycle soya genotypes in Rondonópolis, MT. **Bioscience Journal,** Uberlândia, v. 30, n. 5, p. 1335-1346, Sept./Oct. 2014

SOUZA, R. S.; FERNANDES, M. S. Nitrogen. In: FERNANDES, M. S. (Ed.). Mineral Nutrition of Plants. **Brazilian Society of Soil Science,** 2006. p. 215-252.

SOUZA, T.V.; SILVEIRA, S.C.; SCALON, J. D.; Trail analysis in the relationship between morphological characteristics of maize and its grain yield. **Journal of Statistics,** UFOP. Vol 3, 2014.

SOUZA, V. Q. DE et al. Variance components and association between com hybrids morpho-agronomic characters. **Científica,** v. 43, n. 3, p. 246, 8 July 2015.

SPRAGUE, G. F.; EBERHART, S. A. Com breeding. In: SPRAGUE, W. F. (Ed.). Com and Com Improvement. **Madison: Journal of American Society of Agronomy,** p. 335-336, 1977.

SPRAGUE, G. F.; TATUM, L. A. General and specific combining ability in single crosses of com. **Journal of American Society of Agronomy,** Madison, v.34, n.10, p.923-932,1942.

STEEL, R.G.D.; TORRIE, J.H.; DICKEY, D.A. **Principies and procedures of statistics: a biometrical approach.** 3.ed. New York: McGraw Hill Book, 1997. 666p.

STEVENSON, F. J. Nitrogen agriculture soils. **Madison: Soil Science Society of America,** 1982, p.605-649.

STRECK EV, KÀMPF N, DALMOLIN RSD, KLAMT E, NASCIMENTO PC, SCHNEIDER P, GIASSON E, PINTO LFS. **Soils of Rio Grande do Sul,** 2ª ed, Porto Alegre, EMATER/RS-ASCAR, 222p, 2008.

STRECK, N. A.; WEISS, A.; XUE, Q.; BAENZIGER, P. S. Incorporating a chronology response into the prediction of leaf appearance rate in winter wheat, **Annals of Botany,** v.92, n.2, p.181-190, 2003.

SUBEDI, K. D.; MA, B. L.; XUE, A. G. 2OO7.Planting date and nitrogen effect songrain yield and protein content of spring wheat. **Crop Science,** v. 47, n. 1, p. 36-44.

SUPRAYOGI, Y.; CLARKE, J. M.; BUECKERT, R.; CLARKE, F. R.; POZNIAK, C. J. 2011.Nitrogenremobilisationand post-anthesis nitrogen uptake in relationto elevated grain protein concentration in durum wheat. **Canadian Journal of Plant Science,** v. 91, n. 2, p. 273-282.

SÁ, M.E.; ARF, O.; RODRIGUES, R.A.F.; OLIVEIRA, G.S. 1997.Effects of sowing times on the production and physiological quality of seeds of nine cultivars of sprinkler irrigated rice. **Revista Brasileira de Sementes,** v.19, n.2, p.244-253.

TA, C.T.; WEILAND, R.T. Nitrogen partitioning in maize during ear evelopment. **Crop Science,** v.32, p.443-451, 1992.

TAIZ, L. ZEIGER, E. **Plant Physiology.** 4. ed., 842 p., Artmed, Porto Alegre, 2009.

TAIZ, L.; ZEIGER, E. **Plant physiology.** 3.ed. Porto Alegre: Artmed, 2004, 613p.

TASCA, F.A.; ERNANI. P.R.; ROGERI, D.A.; GATIBONI, L.C.; CASSOL, P.C. Soil ammonia volatilisation after application of conventional urea or urea with a urease inhibitor. **Revista Brasileira de Ciência do Solo, v.** 35, p. 493-502, 2011.

TAVARES, M.; MELO, A. M. T.; SOVITTARO, W. B. Direct and indirect effects and canonical correlations for yield-related characters in chilli peppers. **Bragantia,** Campinas, v.58, n.l, p.41-47,1999.

TEIXEIRA FILHO, M. C. M.; BUZETTI, S.; ALVARES, R. C. F.; FREITAS, J. G.; ARF, O.; SÁ, M. E. Agronomic performance of wheat cultivars in response to plant population and nitrogen fertilisation. **Científica,** Jaboticabal, v.26, n.2, p.97-106, 2008.

TEIXEIRA, F. F.; SOUZA, B. O.; ANDRADE, R. V.; PADILHA, L.Boas Práticas na Manutenção de Germoplasma e Variedades Crioulas de Milho. Sete Lagoas - MG. **Technical Communication,** n. 113, p. 1-8, 2005.

TERRON, A.; PRECIADO, E.; CÓRDOVA, H.; MICKELSON, H.; LOPEZ, R. Determination of the heterotic sire of 30 maize lines derived from lapoblacion 43 SR of CIMMYT. **Agronomia Mesoamericana,** v. 8, p. 26-34,1997.

THOMAS, A. L.; COSTA, J. A. **Soya: management for high grain productivity.** Porto Alegre: Evangraf, 2010. 248p.

TOEBE, M.; CARGNELUTTI FILHO, A. Multivariate non-normality and multicollinearity in path analysis in maize. **Pesquisa Agropecuária Brasileira.** Brasília, DF, v. 48, n. 5, p. 466-477, May 2013.

TOMASINI, R. G. A. & AMBROSI, I. Aspectos económicos da cultura de trigo,

Brasília, v.15, n.2, p.59-84,1998.

TRENKEL, M.E. Slow- and Controlled-Release and Stabilized Fertilizers: An Option for Enhancing Nutrient Use Effciency in Agriculture.International Fertilizer Industry Association (IFA). Paris, France, 2010.

TROYER AF, WELLIN EJ. Heterosis decreasing in hybrids: yield test inbreeds. **Crop Science,** Madison, v. 49, n. 6, p. 1969-1976, 2009.

TROYER, A. F. Adaptedness and Heterosis in Com and Mule Hybrids. **Crop Science,** v. 46, n. 2, p. 528-543, 2006.

UNDERWOOD, E. J.; SUTTLE, N. F. **Mineral nutrition of livestock.** 3. ed. London: CAB International, 614 p.,1999.

USDA - United States **Agency for International Development.** Available at: <http:// http://www.usda.gov>. Accessed on: 06 December 2018.

VALADÃO, F. C.; WEBER, O, L. S.; VALADÃO JÚNIOR, D. D.; SCAPINELLI. A.; DEINA,F. R.; BIANCHINI. A.Phosphate fertilisation and soil compaction: soya and maize root system and soil physical attributes. **Revista Brasileira de Ciência do Solo,** v. 39, n. 1, p. 243- 255, 2015.

VALÉRIO, I. P.; CARVALHO, F, I. F.; BENIN.G.; DA SILVEIRA.G,; SILVA, J, A. G.; NORNBERG.R.; Hagemann.T.;LOUCHE, H. S.; OLIVEIRA, A. C. Seeding density in wheat: the more, the merrier? **Scientia Agrícola,** Piracicaba, v. 70, n. 3, p. 176-184,2013.

VALÉRIO, I. P.; CARVALHO, F. I. F.; OLIVEIRA, A. C.; BENIN, G.; MAIA, L. C.; SILVA, J. A. G.; SCHMIDT, D. M.; SILVEIRA, G. Factors related to the production and development of offspring in wheat. **Semina,** Londrina, v.30, n.l, p. 1207-1218, 2009.

VALÉRIO, I. P.; CARVALHO, F. I. F.; OLIVEIRA, A. C.; LORENCETTI, C.; SOUZA, V. Q; SILVA, J. A. G.; HARWING, L; SCHMIDT, A. M.; BERTAN, L; RIBEIRO, G. Stability of production and combining ability of different oat populations. **Semina: Ciências Agrárias,** Londrina, v. 30, n. 2, p. 331-346, Apr./Jun. 2009.

VALÉRIO, I. P.; CARVALHO, F. I. F.; OLIVEIRA, A. C.; MACHADO, A. A.; BENIN, G.; SCHEEREN, P. L.; SOUZA, V. Q.; HARTWING, I. Offspring development and yield components in wheat genotypes under different sowing densities. **Pesquisa Agropecuária Brasileira,** Brasília, v. 43, n.3,p. 319-326, 2008.

VAN SOEST, P.J. **Nutritional ecology of the ruminant.** Comell University Press. 2.ed. 1994. p.476.

VEIGA, R. D.; FERREIRA, D. F.; RAMALHO, M. A. P. Efficiency of circulating diallels in the choice of genitors. **Pesquisa Agropecuária Brasileira,** Brasília, v. 35, n. 7, p. 1395-1406, jul. 2000.

VELHO, João Pedro et al. Bromatological composition of maize silage produced at different compaction densities. **Revista Brasileira de Zootecnia,** v.36, n.5, p. 1532-1538, 2007.

VENCOVSKY, R. Quantitative inheritance. In: PATERNIANI, E.; VIEGAS, G. P. (Ed.). Maize improvement and production. **Campinas: Cargill Foundation,** 1987. p. 137-209.

VENCOVSKY, R.; BARRIGA, P. **Biometric genetics in plant breeding.** Ribeirão Preto: Brazilian Genetics Society, 1992. 496 p.

VERNETTI, F. J. **Soya: plant, climate, pests, diseases and invaders.** Campinas: Cargill Foundation, 1983. lOOp.

VESOHOSKI, F.; MARCHIORO, V. S.; FRANCO, F. A.; CANTELLE, A. Wheat grain yield components and their direct and indirect effects on productivity. **Revista Ceres,** Viçosa, v. 58, n.l, p. 337-341, 2011.

VIANA, E.M.; KTEHL, R.D.C. **Interaction of nitrogen and potassium on nutrition, chlorophyll content and nitrate reductase activity in wheat plants.** Master's dissertation - Luiz de Queiroz College of Agriculture. 2007.

VICTÓRIA, R.L.; PICCOLO, M.C.; VARGAS, A.A.T. The nitrogen cycle. In; CARDOSO, E.J.B.N.; TSAI, S.M.; NEVES, M.C.P. Microbiologia do solo. **Campinas: SBSC,** 1992. P 105-119.

VIEGAS, G. P. and MIRANDA FILHO, J. B. 1978. Hybrid maize. In: Patemian (ed.) Maize Improvement and Production in Brazil. Cargil Foundation Edition, p. 257 - 309.

VIEIRA, R.D; FORNASIERI FILHO, D.; MINOHARA, L; BERGAMASCHI, M.C.M. Efeito de doses e de 'épocas de aplicação de nitrogénio em cobertura na produção e na qualidade fisiológica de sementes de trigo. **Cientifica,** São Paulo, v. 23, n.2, p. 257-264,1995.

VIEIRA, V.C.; MARTIN, T.N.; MENEZES, L.F.G.; ORTIZ, S.; BERTONCELLI, P.; STORCK, L. Bromatological characterisation of maize silages from super early genotypes. **Ciência Rural,** Santa Maria, v.43, n.ll, p.1925-1931, nov, 2013.

VIGANÓ, L; BRACCINI, A. DE L. E; SCAPIM, C. A.; FRANCO, F. DE A.; SCHUSTER, L; MOTERLE, L. M.; TEXEIRA, L. R. 2010.Physiological quality of wheat seeds in response to the effects of sowing years and seasons. **Revista Brasileira de Sementes,** v. 32, n. 3, p. 86-96.

VILELA, D. Fodder conservation system, 1) silage. Coronel Pacheco: Embrapa-

CNPGL, **Embrapa-CNPGL. Research Bulletin, 11,** 1985. 15 p.

VOLP, A.C.P.; RENHE, I.R.T.; STRINGUETA, P.C. Bioactive natural pigments. **Alimentos e Nutrição Araraquara,** v. 20, n. 1, p. 157-166, 2009.

WALTER, L. C.; STRECK, N. A.; ROSA, H. T.; ALBERTO, C. M.; OLIVEIRA, F. B. Vegetative and reproductive development of wheat cultivars and its association with leaf emission. **Ciência Rural,** Santa Maria, v.39, n.8, p.2320-2326, 2009.

WENDT, W.; CAETANO, V. R. GARCIA, C. A. N. Management of Wheat Crops for Dual Purpose-Forage and Grain. **Pelotas-RS,** n.141,2006, 15p.

WIESER, H.; HARTMANN, G.; KOEHLER, P. Studies on the degradation of gluten proteins during germination of wheat. In: INTERNATIONAL GLUTEN WORKSHOP, 9., 2006, San Francisco, CA, USA.

WRIGHT, S. Correlation and causation. **Journal Agriculture Research,** Washington, v.20, p.557-585, 1921.

WRIGHT, S. The theory of path coefficients: a replay to Niles' criticism. Genetics, Austin, v. 8, n. 3, p. 239-255, 1923. **Campinas: Cargill Foundation** 137-209,1987.

WRIGHT, S. The theory of path coefficients: a replay to Niles' criticism. **Genetics,** Austin, v.8, n.3, p. 239-255, 1923.

WRIGLEY, C.W. Developing better strategies to improve grain quality for wheat. **Australian Journal of Agricultural Research,** Melboume, v.45, p.1-17, 1994.

Wendt, W., CAETANO, V. D. R., & Nunes, C. D. M. (2007). Grain yield and production factors of wheat as a function of rainfall during the reproductive phase. **Embrapa Clima Temperado.**

YAN, W.; HOLLAND, J. B.2010.A heritability-adjusted GGE biplot for test environment evaluation. **Euphytica,** v. 171, n. 3, p. 355-369.

YAN, W.; HOLLAND, J. B.A Heritability-adjusted GGE biplot for test environment evaluation. **Euphytica,** Wageningen, v. 171, n. 3, p. 355-369, 2010.

YANO, G.T.; TAKAHASHI, H.W.; WATANABE, T.S. Evaluation of nitrogen sources and cover application times for wheat cultivation. **Semina: Ciências Agrárias,** Londrina, v. 26, n. 2, p. 141-148, Apr./Jun. 2005.

ZAGO, C. P Maize and sorghum hybrids for silage: agronomic and nutritional characteristics. In: SYMPOSIUM ON STRATEGIC PASTURE MANAGEMENT, 1, 2002, Viçosa. **Proceedings...** Viçosa: UFV, 2002. p. 351-372.

ZARDO, F.P. **Laboratory analyses for the quality control of wheat flour.** Final paper for the Food Technology course at the Federal Institute of Science and Technology of Rio Grande do Sul - Bento Gonçalves Campus. 46 p., 2010.

ZENI-NETO, H. et al. Selection for productivity, stability and adaptability of sugarcane clones in three environments in the state of paraná via mixed models. **Scientia agraria,** v. 9, n. 4, p. 425^430, 2008.

ZHOU, Y.; ZHANG, Y.; WANG, X.; CUI, J.; XIA, X.; SHI, K.; YU, J. Effects of nitrogen form on growth, CO_2 assimilation, chlorophyll fluorescence, and photosynthetic electron allocation in cucumber and rice plants. **Journal of Zhejiang University-Science B, Hangzhou.** v.12, n. 2, p. 126-134, 2011

Printed by Books on Demand GmbH, Norderstedt / Germany